软件项目开发全程实录

Vue.js 项目开发全程实录

明日科技　编著

清华大学出版社
北京

内 容 简 介

本书精选 Vue.js 开发方向的 10 个热门应用项目，实用性非常强。这些项目包含：智汇企业官网首页设计、贪吃蛇小游戏、时光音乐网首页设计、游戏公园博客、电影易购 APP、淘贝电子商城、畅联通讯录、仿饿了么 APP、仿今日头条 APP、四季旅游信息网。本书从软件工程的角度出发，按照项目开发的顺序，系统而全面地讲解每一个项目的开发实现过程。体例上，每章聚焦一个项目，统一采用"开发背景→系统设计→技术准备→各功能模块的设计与实现→项目运行→源码下载"的形式完整呈现项目。这样的安排旨在让读者在学习过程中获得清晰的成就感，并帮助读者快速积累实际项目经验与技巧，以早日实现就业目标。

另外，本书还配备了丰富的 Web 前端在线开发资源库和电子课件，主要内容如下：

- ☑ 技术资源库：439 个核心技术点
- ☑ 项目资源库：13 个精选项目
- ☑ 视频资源库：677 集学习视频
- ☑ 实例资源库：393 个应用实例
- ☑ 源码资源库：406 套项目与案例源码
- ☑ PPT 电子课件

本书可为 Vue.js 入门自学者提供更广泛的项目实战场景，可为计算机专业学生进行项目实训、毕业设计提供项目参考，可供计算机专业教师、IT 培训讲师用作教学参考资料，还可作为前端工程师、IT 求职者、编程爱好者进行项目开发时的参考书。

本书封面贴有清华大学出版社防伪标签，无标签者不得销售。
版权所有，侵权必究。举报：010-62782989，beiqinquan@tup.tsinghua.edu.cn。

图书在版编目（CIP）数据

Vue.js 项目开发全程实录 / 明日科技编著. -- 北京：清华大学出版社，2024.9. --（软件项目开发全程实录）.
ISBN 978-7-302-66854-1

Ⅰ. TP393.092.2

中国国家版本馆 CIP 数据核字第 2024X1R047 号

责任编辑：贾小红
封面设计：秦　丽
版式设计：文森时代
责任校对：马军令
责任印制：沈　露

出版发行：清华大学出版社
网　　址：https://www.tup.com.cn，https://www.wqxuetang.com
地　　址：北京清华大学学研大厦 A 座　　　　邮　编：100084
社 总 机：010-83470000　　　　　　　　　　邮　购：010-62786544
投稿与读者服务：010-62776969，c-service@tup.tsinghua.edu.cn
质量反馈：010-62772015，zhiliang@tup.tsinghua.edu.cn
印 装 者：北京鑫海金澳胶印有限公司
经　　销：全国新华书店
开　　本：203mm×260mm　　　印　张：19.75　　　字　数：640 千字
版　　次：2024 年 9 月第 1 版　　　　　　　　印　次：2024 年 9 月第 1 次印刷
定　　价：89.80 元

产品编号：107420-01

如何使用本书开发资源库

本书赠送价值 999 元的"Web 前端在线开发资源库"一年的免费使用权限。结合图书和开发资源库，读者可以快速提升编程水平，并增强解决实际问题的能力。

1．VIP 会员注册

读者可以刮开图书封底的防盗码并进行扫描，按提示绑定手机微信，然后扫描右侧的二维码，打开明日科技账号注册页面。填写完注册信息后，读者将自动获得一年（自注册之日起）的 Web 前端在线开发资源库的 VIP 使用权限。

Web 前端开发资源库

读者在注册、使用开发资源库时有任何问题，均可通过明日科技官网页面上的客服电话进行咨询。

2．开发资源库简介

Web 前端开发资源库提供了技术资源库（439 个核心技术点）、实例资源库（393 个应用实例）、项目资源库（13 个精选项目）、源码资源库（406 套项目与案例源码）、视频资源库（677 集学习视频），共计五大类、1928 项学习资源。学会、练熟、用好这些资源，读者可在短时间内快速提升自己的开发水平，从一名新手晋升为一名软件工程师。

3．开发资源库的使用方法

在学习本书的各项目时，读者可以利用 Web 前端开发资源库提供的大量技术点、技巧、热点实例、视频等内容，快速回顾或学习相关的知识和技能，从而提高学习效率。

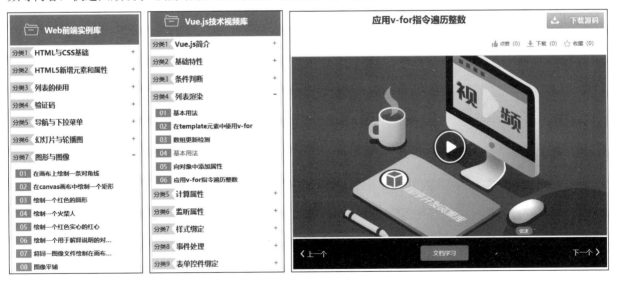

除此之外，开发资源库还提供了更多的大型实战项目，供读者进一步扩展学习，增强编程兴趣和信心，同时积累丰富的项目经验。

此外，读者还可以使用页面上方的搜索栏，快速查阅技术、技巧、实例、项目、源码、视频等资源。

当一切准备就绪，读者便可以踏入软件开发的主战场，接受实战的考验。本书资源包中提供了 Web 前端企业面试真题，是求职面试的绝佳指南，可扫描图书封底的"文泉云盘"二维码获取。

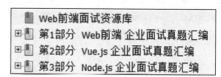

前 言
Preface

丛书说明："软件项目开发全程实录"丛书第 1 版于 2008 年 6 月出版，因其定位于项目开发案例、面向实际开发应用，并解决了社会需求和高校课程设置相对脱节的痛点，在软件项目开发类图书市场上产生了很大的反响，在全国软件项目开发零售图书排行榜中名列前茅。

"软件项目开发全程实录"丛书第 2 版于 2011 年 1 月出版，第 3 版于 2013 年 10 月出版，第 4 版于 2018 年 5 月出版。经过十六年的锤炼打造，不仅深受广大程序员的喜爱，还被百余所高校选为计算机科学、软件工程等相关专业的教材及教学参考用书，更被广大高校学子用作毕业设计和工作实习的必备参考用书。

"软件项目开发全程实录"丛书第 5 版在继承前 4 版所有优点的基础上，进行了大幅度的改版升级。首先，结合当前技术发展的最新趋势与市场需求，增加了程序员求职急需的新图书品种；其次，对图书内容进行了深度更新、优化，新增了当前热门的流行项目，优化了原有经典项目，将开发环境和工具更新为目前的新版本等，使之更与时代接轨，更适合读者学习；最后，录制了全新的项目精讲视频，并配备了更加丰富的学习资源与服务，可以给读者带来更好的项目学习及使用体验。

Vue.js 是一套用于构建用户界面的渐进式 JavaScript 框架。它是一个成熟的、经历了无数实践考验的框架，也是目前使用最广泛的前端框架之一。Vue.js 可以轻松地应对大多数常用场景下的 Web 前端构建，几乎不需要手动优化，并且完全有能力处理大规模的应用。本书以中小型项目为载体，带领读者亲身体验软件开发的实际过程，可以让读者深刻体会 Vue.js 核心技术在项目开发中的具体应用。全书内容不是枯燥的语法和陌生的术语，而是一步一步地引导读者实现一个个热门的项目，从而激发读者学习软件开发的兴趣，将被动学习转变为主动学习。另外，本书的项目开发过程完整，不仅可以为 Web 前端自学者提供项目开发参考，还可以作为大学生毕业设计的项目参考用书。

本书内容

本书提供 Vue.js 开发方向的 10 个热门应用项目，涉及企业门户类、电商购物类、外卖点餐类、信息流类、游戏平台类、旅游信息类等前端开发的多个重点应用方向。具体项目包括：智汇企业官网首页设计、贪吃蛇小游戏、时光音乐网首页设计、游戏公园博客、电影易购 APP、淘贝电子商城、畅联通讯录、仿饿了么 APP、仿今日头条 APP、四季旅游信息网。

本书特点

- ☑ **项目典型**。本书精选了 10 个当前实际开发领域中常见的热点项目，并从实际应用角度出发，对每个项目进行了系统性的讲解，旨在帮助读者积累丰富的开发经验。
- ☑ **流程清晰**。本书项目从软件工程的角度出发，统一采用"开发背景→系统设计→技术准备→各功能模块的设计与实现→项目运行→源码下载"的形式呈现项目内容，这样的布局旨在让读者更全面地了解项目的完整开发流程，从而增强读者的成就感与自信心。
- ☑ **技术新颖**。本书所有项目的实现技术均基于目前业内推荐的最新稳定版本，确保了技术的先进性和实用性。同时，每个项目均设有"技术准备"小节，其中对 Vue.js 基本技术点、高级应用以及第三方组件库等进行了详尽的讲解，这在 Vue.js 基础知识和项目开发之间搭建了一座有效的桥梁，为仅有 Vue.js 基础的初级编程人员参与项目开发提供了清晰的路径，扫除了障碍。

- ☑ **精彩栏目**。本书根据项目学习的实际需求，在每个项目讲解的关键环节增设了"注意""说明"等特色栏目，旨在精准点拨项目的开发要点和精髓，以帮助读者更快地掌握相关技术的应用技巧。
- ☑ **源码下载**。本书每个项目最后都安排了"源码下载"小节，读者可以通过扫描相应二维码下载项目的完整源代码，便于学习和参考。
- ☑ **项目视频**。本书为每个项目都配备了开发、使用指导的精讲微视频，旨在帮助读者更加轻松地搭建、运行和使用项目，并便于读者随时随地进行查看和学习。

读者对象

- ☑ 初学编程的自学者
- ☑ 参与项目实训的学生
- ☑ 做毕业设计的学生
- ☑ 参加实习的初级程序员
- ☑ 高等院校的教师
- ☑ IT 培训机构的教师与学员
- ☑ 程序测试及维护人员
- ☑ 编程爱好者

资源与服务

本书提供了大量的辅助学习资源，同时还提供了专业的知识拓展与答疑服务，旨在帮助读者提高学习效率并解决学习过程中遇到的各种疑难问题。读者需要刮开图书封底的防盗码（刮刮卡），扫描并绑定微信，获取学习权限。

- ☑ **开发环境搭建视频**

搭建环境对于项目开发非常重要，它确保了项目开发在一致的环境下进行，减少了因环境差异导致的错误和冲突。通过搭建开发环境，可以方便地管理项目依赖，提高开发效率。本书提供了开发环境搭建讲解视频，可以引导读者快速准确地搭建本书项目的开发环境。扫描右侧二维码即可观看学习。

开发环境
搭建视频

- ☑ **项目精讲视频**

本书每个项目均配有对应的项目精讲微视频，主要针对项目的需求背景、应用价值、功能结构、业务流程、实现逻辑以及所用到的核心技术点进行精要讲解，可以帮助读者了解项目概要，把握项目要领，快速进入学习状态。扫描每章首页的对应二维码即可观看学习。

- ☑ **项目源码**

本书每章一个项目，系统全面地讲解了该项目的设计及实现过程。为了方便读者学习，本书提供了完整的项目源码（包含项目中用到的所有素材，如图片、数据表等）。扫描每章最后的二维码即可下载。

- ☑ **AI 辅助开发手册**

在人工智能浪潮的席卷之下，AI 大模型工具呈现百花齐放之态，辅助编程开发的代码助手类工具不断涌现，可为开发人员提供技术点问答、代码查错、辅助开发等非常实用的服务，极大地提高了编程学习和开发效率。为了帮助读者快速熟悉并使用这些工具，本书专门精心配备了电子版的《AI 辅助开发手册》，不仅为读者提供各个主流大语言模型的使用指南，而且详细讲解文心快码（Baidu Comate）、通义灵码、腾讯云 AI 代码助手、iFlyCode 等专业的智能代码助手的使用方法。扫描右侧二维码即可阅读学习。

AI 辅助
开发手册

☑ **代码查错器**

为了进一步帮助读者提升学习效率，培养良好的编码习惯，本书配备了由明日科技自主开发的代码查错器。读者可以将本书的项目源码保存为对应的 txt 文件，存放到代码查错器的对应文件夹中，然后自己编写相应的实现代码并与项目源码进行比对，快速找出自己编写的代码与源码不一致或者发生错误的地方。代码查错器配有详细的使用说明文档，扫描右侧二维码即可下载。

代码查错器

☑ **Web 前端开发资源库**

本书配备了强大的线上 Web 前端开发资源库，包括技术资源库、实例资源库、项目资源库、源码资源库、视频资源库。扫描右侧二维码，可登录明日科技网站，获取 Web 前端开发资源库一年的免费使用权限。

Web 前端
开发资源库

☑ **Web 前端面试资源库**

本书配备了 Web 前端面试资源库，精心汇编了大量企业面试真题，是求职面试的绝佳指南。扫描本书封底的"文泉云盘"二维码即可获取。

☑ **教学 PPT**

本书配备了精美的教学 PPT，可供高校教师和培训机构讲师备课使用，也可供读者做知识梳理。扫描本书封底的"文泉云盘"二维码即可下载。另外，登录清华大学出版社网站（www.tup.com.cn），可在本书对应页面查阅教学 PPT 的获取方式。

☑ **学习答疑**

在学习过程中，读者难免会遇到各种疑难问题。本书配有完善的新媒体学习矩阵，包括 IT 今日热榜（实时提供最新技术热点）、微信公众号、学习交流群、400 电话等，可为读者提供专业的知识拓展与答疑服务。扫描右侧二维码，根据提示操作，即可享受答疑服务。

学习答疑

致读者

本书由明日科技前端开发团队组织编写，主要编写人员有张鑫、王小科、王国辉、刘书娟、赵宁、高春艳、赛奎春、田旭、葛忠月、杨丽、李颖、程瑞红、张颖鹤等。明日科技是一家专业从事软件开发、教育培训以及软件开发教育资源整合的高科技公司，其编写的图书非常注重选取软件开发中的必需、常用内容，同时很注重内容的易学性、学习的方便性以及相关知识的拓展性，深受读者喜爱。其编写的图书多次荣获"全行业优秀畅销品种""全国高校出版社优秀畅销书"等奖项，多个品种长期位居同类图书销售排行榜的前列。

在编写本书的过程中，我们始终本着科学、严谨的态度，力求做到精益求精。然而，书中难免存在疏漏和不妥之处，我们在此诚挚欢迎读者提出任何意见和建议。

感谢您选择本书，希望本书能成为您的良师益友，成为您步入编程高手之路的踏脚石。

宝剑锋从磨砺出，梅花香自苦寒来。祝读书快乐！

编　者
2024 年 8 月

目 录

第 1 章　智汇企业官网首页设计 1
——事件处理 + 表单元素绑定 + 样式绑定 + CSS 过渡

- 1.1 开发背景 1
- 1.2 系统设计 2
  - 1.2.1 开发环境 2
  - 1.2.2 业务流程 2
  - 1.2.3 功能结构 2
- 1.3 技术准备 2
- 1.4 功能设计 5
  - 1.4.1 导航栏的设计 5
  - 1.4.2 活动图片展示界面 7
  - 1.4.3 企业新闻展示界面 9
  - 1.4.4 产品推荐界面 11
  - 1.4.5 浮动窗口设计 13
- 1.5 项目运行 14
- 1.6 源码下载 15

第 2 章　贪吃蛇小游戏 16
——v-show 指令 + 事件处理 + 表单元素绑定

- 2.1 开发背景 16
- 2.2 系统设计 17
  - 2.2.1 开发环境 17
  - 2.2.2 业务流程 17
  - 2.2.3 功能结构 17
- 2.3 技术准备 18
- 2.4 游戏初始界面设计 19
  - 2.4.1 创建主页 19
  - 2.4.2 游戏初始化 21
  - 2.4.3 设置游戏速度 22
- 2.5 游戏操作 22
  - 2.5.1 键盘按键控制 22
  - 2.5.2 蛇的移动 23

- 2.5.3 游戏结束 24
- 2.6 项目运行 24
- 2.7 源码下载 25

第 3 章　时光音乐网首页设计 26
——Vue CLI + axios

- 3.1 开发背景 26
- 3.2 系统设计 27
  - 3.2.1 开发环境 27
  - 3.2.2 业务流程 27
  - 3.2.3 功能结构 27
- 3.3 技术准备 27
- 3.4 功能设计 29
  - 3.4.1 导航栏的设计 29
  - 3.4.2 歌曲列表展示界面 31
  - 3.4.3 轮播图的设计 33
  - 3.4.4 歌曲排行榜 35
  - 3.4.5 最新音乐资讯 38
  - 3.4.6 新歌首发 40
  - 3.4.7 首页底部的设计 44
  - 3.4.8 在根组件中构建音乐网首页 44
- 3.5 项目运行 45
- 3.6 源码下载 46

第 4 章　游戏公园博客 47
——Vue CLI + Vue Router + Vuex

- 4.1 开发背景 47
- 4.2 系统设计 48
  - 4.2.1 开发环境 48
  - 4.2.2 业务流程 48
  - 4.2.3 功能结构 48
- 4.3 技术准备 48
- 4.4 创建项目 49
- 4.5 功能设计 50
  - 4.5.1 主页设计 50

| 4.5.2 博客列表页面设计 57
| 4.5.3 博客详情页面设计 59
| 4.5.4 关于我们页面设计 62
| 4.5.5 路由配置 65
| 4.6 项目运行 66
| 4.7 源码下载 67

第 5 章 电影易购 APP 68
——Vue CLI + Vue Router + Vuex + axios

| 5.1 开发背景 68
| 5.2 系统设计 69
| 5.2.1 开发环境 69
| 5.2.2 业务流程 69
| 5.2.3 功能结构 69
| 5.3 技术准备 70
| 5.4 创建项目 70
| 5.5 公共组件设计 70
| 5.5.1 头部组件设计 71
| 5.5.2 底部导航栏组件设计 71
| 5.6 影片页面设计 73
| 5.6.1 正在热映影片组件设计 73
| 5.6.2 即将上映影片组件设计 76
| 5.6.3 影片搜索组件设计 78
| 5.6.4 影片页面组件设计 81
| 5.7 选择城市页面设计 84
| 5.8 影院页面设计 90
| 5.8.1 影院列表组件设计 91
| 5.8.2 影院页面组件设计 93
| 5.9 我的页面设计 94
| 5.9.1 用户登录组件设计 94
| 5.9.2 用户注册组件设计 97
| 5.9.3 用户订单和服务组件设计 99
| 5.9.4 我的页面组件设计 102
| 5.10 路由配置 102
| 5.11 项目运行 104
| 5.12 源码下载 104

第 6 章 淘贝电子商城 105
——Vue CLI + Vue Router + Vuex + localStorage

| 6.1 开发背景 105

| 6.2 系统设计 106
| 6.2.1 开发环境 106
| 6.2.2 业务流程 106
| 6.2.3 功能结构 106
| 6.3 技术准备 107
| 6.4 主页的设计与实现 108
| 6.4.1 主页的设计 108
| 6.4.2 顶部区和底部区功能的实现 108
| 6.4.3 商品分类导航功能的实现 112
| 6.4.4 轮播图功能的实现 114
| 6.4.5 商品推荐功能的实现 115
| 6.5 商品详情页面的设计与实现 117
| 6.5.1 商品详情页面的设计 117
| 6.5.2 图片放大镜效果的实现 119
| 6.5.3 商品概要功能的实现 120
| 6.5.4 猜你喜欢功能的实现 123
| 6.5.5 选项卡切换效果的实现 125
| 6.6 购物车页面的设计与实现 127
| 6.6.1 购物车页面的设计 127
| 6.6.2 购物车页面的实现 127
| 6.7 付款页面的设计与实现 129
| 6.7.1 付款页面的设计 129
| 6.7.2 付款页面的实现 130
| 6.8 注册和登录页面的设计与实现 133
| 6.8.1 注册和登录页面的设计 133
| 6.8.2 注册页面的实现 134
| 6.8.3 登录页面的实现 136
| 6.9 项目运行 138
| 6.10 源码下载 139

第 7 章 畅联通讯录 140
——Vue CLI + Vue Router + Vuex + localStorage + sessionStorage

| 7.1 开发背景 140
| 7.2 系统设计 141
| 7.2.1 开发环境 141
| 7.2.2 业务流程 141
| 7.2.3 功能结构 142
| 7.3 技术准备 142
| 7.4 创建项目 143
| 7.5 注册和登录页面设计 144

  7.5.1 页面头部组件设计 145

  7.5.2 用户注册组件设计 146

  7.5.3 用户登录组件设计 149

 7.6 通讯录页面设计 152

  7.6.1 通讯录页面组件设计 152

  7.6.2 通讯录列表组件设计 157

  7.6.3 分页组件设计 160

  7.6.4 联系人组件设计 162

 7.7 添加联系人组件设计 164

 7.8 个人中心组件设计 168

 7.9 路由配置 173

 7.10 项目运行 174

 7.11 源码下载 175

第8章 仿饿了么APP 176

——Vue CLI + Router + axios + JSON Server + localStorage + SessionStorage

 8.1 开发背景 176

 8.2 系统设计 177

  8.2.1 开发环境 177

  8.2.2 业务流程 177

  8.2.3 功能结构 177

 8.3 技术准备 178

 8.4 首页的设计与实现 180

  8.4.1 商家分类页面设计 180

  8.4.2 推荐商家列表页面设计 182

  8.4.3 底部导航栏的设计 183

 8.5 分类商家列表的设计与实现 185

 8.6 商家详情页面的设计与实现 187

  8.6.1 商家信息页面设计 187

  8.6.2 购物车页面设计 190

 8.7 确认订单页面的设计与实现 192

  8.7.1 确认订单页面设计 192

  8.7.2 新增收货地址页面的设计 194

  8.7.3 地址管理页面的设计 196

 8.8 支付页面的设计与实现 198

 8.9 订单列表页面的设计与实现 201

 8.10 注册和登录页面的设计与实现 203

  8.10.1 注册页面的设计 204

  8.10.2 登录页面的设计 206

 8.11 我的页面的设计与实现 208

 8.12 项目运行 209

 8.13 源码下载 210

第9章 仿今日头条APP 211

——Vue CLI + Router + Vuex + axios + JSON Server + Vant + amfe-flexible + Day.js

 9.1 开发背景 211

 9.2 系统设计 212

  9.2.1 开发环境 212

  9.2.2 业务流程 212

  9.2.3 功能结构 213

 9.3 技术准备 213

  9.3.1 技术概览 213

  9.3.2 Vant 213

  9.3.3 amfe-flexible 219

  9.3.4 Day.js 219

 9.4 创建项目 220

 9.5 新闻列表页面的设计与实现 221

  9.5.1 页面主组件设计 221

  9.5.2 新闻列表组件设计 224

  9.5.3 新闻列表项组件设计 227

  9.5.4 频道管理组件设计 230

  9.5.5 底部导航栏的设计 233

 9.6 新闻搜索功能的设计与实现 234

  9.6.1 搜索组件设计 234

  9.6.2 搜索结果组件设计 236

 9.7 新闻详情页面的设计与实现 238

  9.7.1 新闻内容组件设计 238

  9.7.2 用户评论组件的设计 241

 9.8 注册和登录页面的设计与实现 250

  9.8.1 注册页面的设计 250

  9.8.2 登录页面的设计 252

 9.9 我的页面的设计与实现 254

 9.10 路由配置 256

 9.11 项目运行 257

 9.12 源码下载 258

第10章 四季旅游信息网 259

——Vue CLI + Vue Router + axios + JSON Server + ElementPlus + Day.js

 10.1 开发背景 259

 10.2 系统设计 260

10.2.1	开发环境	260
10.2.2	业务流程	260
10.2.3	功能结构	261

10.3 技术准备 261
- 10.3.1 技术概览 261
- 10.3.2 ElementPlus 261
- 10.3.3 Day.js 中的 add()方法和 format()方法 266

10.4 创建项目 266

10.5 公共组件设计 267
- 10.5.1 页面头部组件设计 267
- 10.5.2 页面底部组件设计 269

10.6 首页设计 269

10.7 热门景点页面设计 273
- 10.7.1 景点列表组件设计 274
- 10.7.2 景点列表项组件设计 276
- 10.7.3 景点详情组件设计 277

10.8 酒店住宿页面设计 279
- 10.8.1 酒店列表组件设计 279
- 10.8.2 酒店列表项组件设计 282
- 10.8.3 酒店搜索结果组件设计 283
- 10.8.4 酒店详情组件设计 284

10.9 门票预订页面设计 286

10.10 游客服务页面设计 290
- 10.10.1 游客服务组件设计 291
- 10.10.2 导游组件设计 292
- 10.10.3 游客须知组件设计 294

10.11 用户中心页面设计 295
- 10.11.1 用户注册组件设计 295
- 10.11.2 用户登录组件设计 298

10.12 路由配置 300

10.13 项目运行 303

10.14 源码下载 304

第 1 章 智汇企业官网首页设计

——事件处理 + 表单元素绑定 + 样式绑定 + CSS 过渡

在网络高速发展的时代,很多企业都有自己的官方网站。企业网站不仅用于向外界展示企业的动态,还能宣传企业的产品。本章将运用 Vue.js 中的事件处理、表单元素绑定、样式绑定和 CSS 过渡等技术实现智汇企业官网首页的设计,旨在展示企业动态。

本项目的核心功能及实现技术如下:

项目微视频

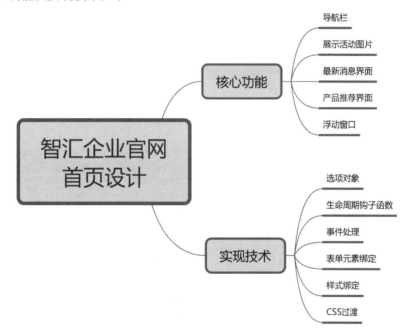

1.1 开发背景

现在很多企业都拥有自己的官方网站,这些官方网站不仅能让更多的用户深入了解公司的文化、业务和产品等,而且在推广公司产品方面发挥着至关重要的作用。此外,企业拥有自己的官方网站还能显著提升用户对公司的信任度。因此,一个企业拥有自己的官方网站是很有必要的。

企业官网开发是 Vue.js 最常见的应用场景之一,本章将使用 Vue.js 开发一个简洁、易操作的"智汇企业官网首页"项目,该项目包含企业官网首页最基本的页面结构和功能,其实现目标如下:

- ☑ 实现活动图片展示界面,以此来展示公司创造的成果和业绩。
- ☑ 以消息滚动的形式显示公司的最新消息。
- ☑ 以图片列表的形式展示公司产品。
- ☑ 通过浮动窗口实现显示在线交流的界面效果。

1.2 系统设计

1.2.1 开发环境

本项目的开发及运行环境如下：
- ☑ 操作系统：推荐 Windows 10、Windows 11 或更高版本，同时兼容 Windows 7（SP1）。
- ☑ 开发工具：WebStorm。
- ☑ 开发框架：Vue.js 3.0。

1.2.2 业务流程

在智汇企业官网首页，网站 logo 下面设计有网站导航栏，导航栏下面则是活动图片展示区域。单击不同的城市名称，可以展示对应城市的活动信息。显示公司的最新消息时，我们采用滚动效果。在产品推荐界面，用户可以对产品进行检索。最后，页面右侧还设计一个浮动窗口。

本项目的业务流程如图 1.1 所示。

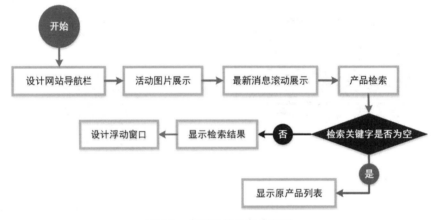

图 1.1 本项目的业务流程图

1.2.3 功能结构

本项目的功能结构已经在章首页中给出。作为企业官网首页的应用，本项目实现的具体功能如下：
- ☑ 浏览活动图片：用户可以通过单击对应的城市名称来浏览该城市的活动图片。
- ☑ 浏览最新消息：最新消息以从下向上滚动的形式进行显示。
- ☑ 展示公司产品：公司产品以图片列表的形式进行展示。
- ☑ 检索产品：用户可以通过输入关键字来检索产品，系统将显示相应的检索结果。
- ☑ 设置浮动窗口：浮动窗口的位置在页面中始终保持不变。

1.3 技术准备

在开发智汇企业官网首页的过程中，我们主要应用 Vue.js 框架。这里将本项目所用的 Vue.js 核心技术点及其具体作用简述如下：

- ☑ 选项对象：每个 Vue.js 的应用都需要使用 createApp()方法创建一个应用程序的实例对象，并将其挂载到指定的 DOM 元素上。createApp()是一个全局 API，它接收一个根组件选项对象作为参数。这个选项对象包括数据、方法等选项。在开发该项目时，我们将需要使用的数据定义在 data 选项中，将实现某个功能的方法定义在 methods 选项中。示例代码如下：

```html
<div id="app">
    <p>{{showInfo()}}</p>
</div>
<script src="https://unpkg.com/vue@3"></script>
<script type="text/javascript">
    //创建应用程序实例
    const vm = Vue.createApp({
        //返回数据对象
        data(){
            return {
                text : '书是人类进步的阶梯。',
                author : ' —— 高尔基'
            }
        },
        methods : {
            showInfo : function(){
                return this.text + this.author;       //连接字符串
            }
        }
    //装载应用程序实例的根组件
    }).mount('#app');
</script>
```

- ☑ 生命周期钩子函数：在创建每个应用程序实例时都会有一系列初始化步骤。例如，创建数据绑定、将实例挂载到 DOM 以及在数据变化时触发 DOM 更新等。在这个过程中，程序会运行一些叫作生命周期钩子的函数，这些钩子函数允许开发者定义业务逻辑。其中，mounted()钩子函数的示例代码如下：

```html
<div id="app">
    <p>当前时间：{{ currentTime }}</p>
</div>
<script src="https://unpkg.com/vue@3"></script>
<script type="text/javascript">
    //创建应用程序实例
    const vm = Vue.createApp({
        //返回数据对象
        data(){
            return {
                currentTime : ''
            }
        },
        mounted(){
            let t = this;
            setInterval(function (){
                t.currentTime = new Date().toLocaleString();
            },1000)
        }
    //装载应用程序实例的根组件
    }).mount('#app');
</script>
```

在开发该项目时，我们将实现最新消息向上滚动效果的方法放置在 mounted()钩子函数中，这样在文档渲染完毕后就可以立即展现最新消息向上滚动的效果。

☑ 事件处理：在 Vue.js 中，监听 DOM 事件需要使用 v-on 指令。v-on 指令后面可以跟任何原生事件名称。通常，事件名称后面会指定一个方法，该方法包含了触发事件时要执行的 JavaScript 代码。示例代码如下：

```
<div id="app">
    <button v-on:click="toggle">{{flag ? '隐藏' : '显示'}}</button>
    <div v-show="flag">{{text}}</div>
</div>
<script src="https://unpkg.com/vue@3"></script>
<script type="text/javascript">
    const vm = Vue.createApp({
        data(){
            return {
                text : '心有多大，世界就有多大！',
                flag : false
            }
        },
        methods : {
            toggle : function(){                        //切换显示状态
                this.flag = !this.flag;
            }
        }
    }).mount('#app');
</script>
```

在该项目中，当用户单击"产品检索"按钮时，会触发 click 事件。随后，该事件会调用 search() 方法，该方法会通过检索关键字对产品列表进行过滤，从而实现检索商品的功能。

☑ 表单元素绑定：在 Vue.js 中，对表单元素进行双向数据绑定需要使用 v-model 指令，在修改表单元素值的同时，Vue 实例中对应的属性值也会随之更新。示例代码如下：

```
<div id="app">
    <textarea rows="6" v-model="text"></textarea>
    <p style="white-space:pre">{{text}}</p>
</div>
<script src="https://unpkg.com/vue@3"></script>
<script type="text/javascript">
    const vm = Vue.createApp({
        data(){
            return {
                text : '两岸猿声啼不住，'
            }
        }
    }).mount('#app');
</script>
```

在该项目中，使用 v-model 指令将文本框和表示检索关键字的 keyword 属性进行绑定，当文本框中的内容发生改变时，绑定的属性值也会随着变化。

☑ 样式绑定：在 Vue.js 中，对元素样式的绑定就是对元素的 class 和 style 属性进行操作。其中，class 属性用于定义元素的类名列表，而 style 属性则用于定义元素的内联样式。要实现对这两个属性进行数据绑定，需要使用 v-bind 指令。示例代码如下：

```
<style>
    .active{
        font-weight:bold;            /*设置字体粗细*/
        font-size:26px;              /*设置字体大小*/
        color:red;                   /*设置文字颜色*/
    }
</style>
```

```
<div id="app">
    <div v-bind:class="{active : isActive}">人生的意义在于付出而不是索取</div>
</div>
<script src="https://unpkg.com/vue@3"></script>
<script type="text/javascript">
    const vm = Vue.createApp({
        data(){
            return {
                isActive : true                              //使用 active 类名
            }
        }
    }).mount('#app');
</script>
```

在该项目中,我们对导航栏中每个菜单项的 class 属性进行样式绑定,以实现当用户单击某个菜单项时,该菜单项会高亮显示的效果。

☑ CSS 过渡:CSS 过渡其实就是一个淡入淡出的效果。当插入或删除包含在<transition>或<transition-group>组件中的元素时可以定义过渡效果。示例代码如下:

```
<style>
    /*设置 CSS 属性名和持续时间*/
    .effect-enter-active, .effect-leave-active{
        transition: opacity 1s
    }
    .effect-enter-from, .effect-leave-to{
        opacity: 0
    }
</style>
<div id="app">
    <button v-on:click="show = !show">{{show ? '隐藏' : '显示'}}</button><br>
    <transition name="effect">
        <p v-if="show">路遥知马力,日久见人心。</p>
    </transition>
</div>
<script src="https://unpkg.com/vue@3"></script>
<script type="text/javascript">
    const vm = Vue.createApp({
        data(){
            return {
                show : true
            }
        }
    }).mount('#app');
</script>
```

在该项目中,使用<transition-group>组件实现切换活动图片时的过渡效果。

《Vue.js 从入门到精通》一书详细地讲解了有关选项对象、生命周期钩子函数、事件处理、表单元素绑定、样式绑定等基础知识,对这些知识不太熟悉的读者,可以参考该书相关章节进行学习。

1.4 功能设计

1.4.1 导航栏的设计

网站首页的导航栏共有 6 个菜单项。除了"首页",还有 5 个菜单项,包括"全部产品""换新服务"

"官方商城""加入智汇"和"商业合作"。导航栏的初始效果如图1.2所示。当用户单击某个导航菜单项时,该菜单项的样式会发生变化,效果如图1.3所示。

图1.2　导航栏的初始效果

图1.3　单击某个菜单项时的效果

设计导航栏的关键步骤如下。

（1）编写 HTML 代码,定义<div>元素,并设置其 id 属性值为 app。在该元素中定义 6 个导航菜单项,使用 v-on 指令对每个菜单项的 click 事件进行监听,再对每个菜单项的 class 属性进行样式绑定。代码如下:

```
<div id="app">
  <div class="cen">
    <div class="menu">
      <span @click="select(1)" :class="{act: tag===1}">首页</span>
      <span @click="select(2)" :class="{act: tag===2}">全部产品</span>
      <span @click="select(3)" :class="{act: tag===3}">换新服务</span>
      <span @click="select(4)" :class="{act: tag===4}">官方商城</span>
      <span @click="select(5)" :class="{act: tag===5}">加入智汇</span>
      <span @click="select(6)" :class="{act: tag===6}">商业合作</span>
    </div>
  </div>
</div>
```

（2）编写 CSS 代码,用于为页面元素设置样式。其中,act 类名定义用户单击导航栏中某个菜单项时的样式。代码如下:

```
<style>
    .menu{
        display:inline-block;              /*设置行内块元素*/
        background-color: #3399FF;         /*设置背景颜色*/
        margin:5px auto;                   /*设置外边距*/
    }
    .menu span{
        display:inline-block;              /*设置行内块元素*/
        width:145px;                       /*设置宽度*/
        height:40px;                       /*设置高度*/
        line-height:40px;                  /*设置行高*/
        cursor:pointer;                    /*设置鼠标光标形状*/
        text-align:center;                 /*设置文本居中显示*/
        font-size: 14px;                   /*设置文字大小*/
        color:#FFFFFF;                     /*设置文字颜色*/
    }
    .act{
        background-color: #9966FF;         /*设置背景颜色*/
        color:#FFFFFF;                     /*设置文字颜色*/
    }
</style>
```

（3）在<script>标签中,我们创建应用程序实例,并定义数据和方法。在 methods 选项中,我们定义 select()方法,该方法会在用户单击某个菜单项时被调用。在 select()方法中,我们将 tag 属性的值设置为传递的参数值,然后通过判断 tag 属性的值来确定是否在菜单项中使用 act 类的样式。代码如下:

```
<script type="text/javascript">
    const vm = Vue.createApp({
```

```
    data(){
      return {
        tag : 1,          //用于控制导航栏是否使用样式
      }
    },
    methods: {
      select : function(value){
        this.tag = value;
      }
    }
  })
</script>
```

> **说明**
> 由于该项目只是实现企业官网的首页，因此导航栏并没有实际意义上的功能。读者如果感兴趣，可以自己设计导航栏菜单项对应的页面。

1.4.2 活动图片展示界面

在智汇企业官网首页中，导航栏下方是企业活动图片展示界面。该界面列出了企业参加的一些地区展会活动的相关信息，包括展会图片、展会名称和图片简介。用户单击左右两侧的图片按钮，即可切换查看不同的展会图片。界面效果分别如图1.4和图1.5所示。

图 1.4　宁波展会活动图片

图 1.5　长春展会活动图片

活动图片展示界面的实现步骤如下。

（1）编写HTML代码，定义<div>元素，并在该元素中使用<transition-group>组件来实现切换展会图片时的过渡效果。在此组件中，我们对展会图片、展会名称和图片简介进行绑定。在<transition-group>组件后，我们定义两个<div>元素，这两个元素用于渲染左右两侧的图片按钮，同时我们对每个图片按钮的class属性进行样式绑定。代码如下：

```
<div class="i02">
  <div class="banner">
    <transition-group name="effect">
```

```
        <div :key="i">
          <div id="ImageCyclerImage"><img :src="info[i].image"></div>
          <div id="ImageCyclerOverlay" class="grey">
            <div id="ImageCyclerOverlayBackground"></div>
            <p class="title">{{info[i].title}}</p>
            <p>{{info[i].desc}}<a href="#">Find out more &gt;</a></p>
          </div>
        </div>
      </transition-group>
      <div id="ImageCyclerTabs">
        <div v-for="(item,index) in leftBanner" :key="index" :id="item.id">
          <a href="#" @click="i = index" :class="{active:i === index}"><img :src="item.url"></a>
        </div>
      </div>
      <div id="Layer1">
        <div v-for="(item,index) in rightBanner" :key="index" :id="item.id">
          <a href="#" @click="i = index + 5" :class="{active:i === index + 5}"><img :src="item.url"></a>
        </div>
      </div>
    </div>
</div>
```

（2）编写 CSS 代码，为元素设置过渡属性，以实现切换展会图片时的过渡效果。代码如下：

```
<style>
  /*设置过渡属性*/
  .effect-enter-active, .effect-leave-active{
    transition: all .5s;
  }
  .effect-enter-from, .effect-leave-to{
    opacity: 0;
  }
</style>
```

（3）在创建的应用程序实例中，定义以下数据：展会图片的索引 i、展会图片信息的列表 info、左侧图片按钮的列表 leftBanner 以及右侧图片按钮的列表 rightBanner。代码如下：

```
<script type="text/javascript">
  const vm = Vue.createApp({
    data(){
      return {
        i: 0,                                    //展会图片的索引
        info: [                                  //展会图片信息的列表
          { image: 'images/hero1.jpg', title: '宁波展会', desc: '消费类电子产品展览中心现场'},
          { image: 'images/hero2.jpg', title: '长春展会', desc: '科技企业高端产品展览中心现场'},
          { image: 'images/hero3.jpg', title: '北京展会', desc: '手机展区新品手机上市一览'},
          { image: 'images/hero4.jpg', title: '大连展会', desc: '华为云计算展示区域一览'},
          { image: 'images/hero5.jpg', title: '戴尔新品上市', desc: '游匣 G16 7630 高性能游戏笔记本'},
          { image: 'images/hero6.jpg', title: '深圳展会', desc: '5G 智能电视展览中心现场'},
          { image: 'images/hero7.jpg', title: '青岛展会', desc: '智能电视展区康佳品牌展览现场'},
          { image: 'images/hero8.jpg', title: '广州展会', desc: '5G 数码电子产品创意展区'},
          { image: 'images/hero9.jpg', title: '南京展会', desc: '华为高端产品展览现场'},
          { image: 'images/hero10.jpg', title: '华为 Mate 60 新品上市', desc: '纵横山海 安心畅联'}
        ],
        leftBanner: [                            //左侧图片按钮的列表
          {id: 'mg', url: 'images/mg.png'},
          {id: 'jnd', url: 'images/jnd.png'},
          {id: 'yg', url: 'images/yg.png'},
          {id: 'dg', url: 'images/dg.png'},
          {id: 'hg', url: 'images/hg.png'}
        ],
        rightBanner: [                           //右侧图片按钮的列表
```

```
            {id: 'fg', url: 'images/fg.png'},
            {id: 'rb', url: 'images/rb.png'},
            {id: 'xjp', url: 'images/xjp.png'},
            {id: 'odly', url: 'images/odly.png'},
            {id: 'qt', url: 'images/qt.png'}
        ]
    }
  }
})
</scrlpt>
```

1.4.3 企业新闻展示界面

企业新闻展示界面主要用于展示企业的新闻列表，这些新闻以从下向上滚动的形式进行展示。界面效果分别如图 1.6 和图 1.7 所示。

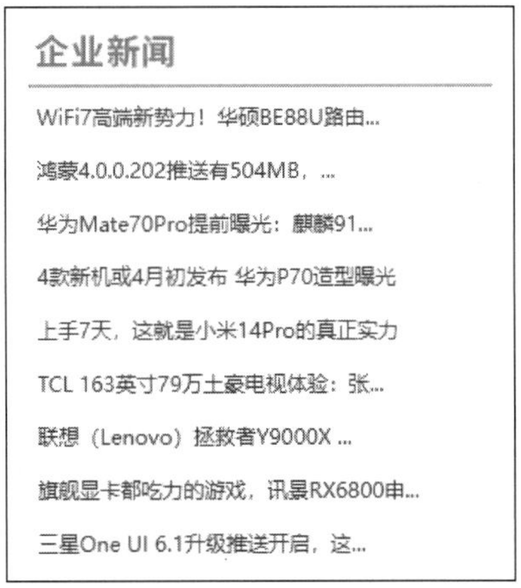

图 1.6　最新消息列表

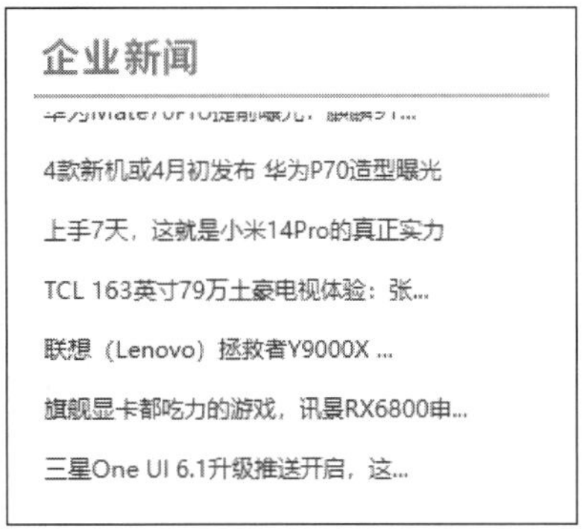

图 1.7　最新消息列表向上滚动

企业新闻展示界面的实现步骤如下。

（1）编写 HTML 代码，定义<div>元素，在该元素中定义 ul 列表，并对该列表的 style 属性进行样式绑定。当触发列表的 mouseenter 事件时，调用 stop()方法；当触发列表的 mouseleave 事件时，调用 up()方法。接着，对 ul 列表中的 li 列表项使用 v-for 指令，对最新消息列表 news_list 进行遍历，并在遍历时调用 subStr()方法对企业新闻标题进行截取。代码如下：

```html
<div class="i03c">
        <div><img src="images/i06.gif"></div>
        <div id="layout">
          <div class="scroll">
            <ul class="list" :style="{top:dis + 'px'}" @mouseenter="stop" @mouseleave="up">
              <li v-for="value in news_list" :key="value">{{subStr(value)}}</li>
            </ul>
          </div>
        </div>
</div>
```

（2）编写 CSS 代码，为 div 元素和 ul 列表设置样式。关键代码如下：

```
<style>
    .scroll{
        margin-left:5px;                            /*设置左外边距*/
        margin-top:5px;                             /*设置上外边距*/
        width:260px;                                /*设置宽度*/
```

```css
      height:300px;                                    /*设置高度*/
      overflow:hidden;                                 /*设置溢出内容隐藏*/
      position: relative;                              /*设置相对定位*/
    }
    .scroll ul{
      position: absolute;                              /*设置绝对定位*/
      top: 0;                                          /*设置顶部距离*/
    }
    .scroll li{
      width:260px;                                     /*设置宽度*/
      height:30px;                                     /*设置高度*/
      line-height:30px;                                /*设置行高*/
    }
</style>
```

（3）在创建的应用程序实例中定义数据、方法和 mounted()钩子函数。scrollUp()方法用于实现最新消息列表向上滚动的效果，stop()方法用于停止向上滚动效果，up()方法用于调用 scrollUp()方法实现列表的向上滚动效果。subStr()方法用于截取消息标题的前 20 个字符。在 mounted()钩子函数中调用 scrollUp()方法，在文档渲染完毕后实现列表的向上滚动效果。代码如下：

```javascript
<script type="text/javascript">
  const vm = Vue.createApp({
    data(){
      return {
        dis:0,                                         //向上滚动距离
        timerID: null,                                 //定时器 ID
        news_list:[                                    //最新消息列表
          "实用折叠仅此一家！vivo X Fold3 全面评测！",
          "WiFi7 高端新势力！华硕 BE88U 路由器开箱",
          "鸿蒙 4.0.0.202 推送有 504MB，内容是这些，你升级了吗？",
          "华为 Mate70Pro 提前曝光：麒麟 9100+极致无边框+顶级昆仑屏",
          "4 款新机或 4 月初发布 华为 P70 造型曝光",
          "上手 7 天，这就是小米 14Pro 的真正实力",
          "TCL 163 英寸 79 万土豪电视体验：张开双臂都没有屏幕宽",
          "联想（Lenovo）拯救者 Y9000X 2024 新品笔记本上市",
          "旗舰显卡都吃力的游戏，讯景 RX6800 申请出战",
          "三星 One UI 6.1 升级推送开启，这些机型率先支持"
        ]
      }
    },
    methods: {
      scrollUp: function (){                           //向上滚动
        var t = this;
        this.timerID = setInterval(function (){
          t.dis = t.dis === -300 ? 0 : t.dis - 0.5;    //设置向上滚动距离
        }, 20);
      },
      stop: function() {
        clearInterval(this.timerID);                   //停止向上滚动操作
      },
      up: function() {
        this.scrollUp();                               //执行向上滚动操作
      },
      subStr: function (value){
        if(value.length > 20){                         //如果标题长度大于 20
          return value.substr(0,20) + '...';           //截取标题前 20 个字符
        }else{
          return value;                                //返回原标题
        }
      },
```

```
    mounted: function (){
      this.scrollUp();                                          //自动执行滚动效果
    }
  })
</script>
```

1.4.4 产品推荐界面

产品推荐界面主要展示了企业的推荐产品列表,并提供了产品检索功能。用户可以在文本框中输入检索关键字,然后单击"产品检索"按钮,界面下方会显示检索到的产品列表。界面效果分别如图 1.8 和图 1.9 所示。

图 1.8　产品推荐列表

图 1.9　检索结果

产品推荐界面的实现步骤如下。

(1)编写 HTML 代码,定义两个<div>元素。其中:在第一个<div>元素中定义一个用于输入检索关键字的文本框和一个图片按钮,使用 v-model 指令将该文本框和 keyword 属性进行绑定,当单击图片按钮时调用 search()方法;在第二个<div>元素中定义一个 ul 列表,在该列表中使用 v-for 指令对检索结果列表 searchResult 进行遍历,在遍历时输出产品图片和产品名称。代码如下:

```
<div class="search">
  <img src="images/i11.gif" @click="search">
  <input type="text" placeholder="请输入搜索关键字" v-model="keyword">
```

```html
</div>
<div class="product">
  <ul>
    <li v-for="item in searchResult">
      <img width="160" :src="item.url">
      <div>{{item.name}}</div>
    </li>
  </ul>
</div>
```

（2）编写 CSS 代码，为<div>元素、文本框和 ul 列表设置样式。关键代码如下：

```css
<style>
.search{
    width: 100%;                                            /*设置宽度*/
    height: 44px;                                           /*设置高度*/
    background-image: url("../images/i10.gif");             /*设置背景图像*/
}
.search input,.search img{
    float: right;                                           /*设置左浮动*/
    margin-top: 8px;                                        /*设置上外边距*/
    margin-right: 3px;                                      /*设置右外边距*/
}
.search input{
    border:solid 1px #CFCECE;                               /*设置边框*/
    width:150px;                                            /*设置宽度*/
    height:18px;                                            /*设置高度*/
}
.search img{
    cursor: pointer;                                        /*设置鼠标形状*/
}
.product{
    width: 100%;                                            /*设置宽度*/
    text-align: center;                                     /*设置文本水平居中显示*/
}
.product ul{
    list-style: none;                                       /*设置列表无样式*/
    margin: 5px auto;                                       /*设置外边距*/
}
.product ul li{
    float: left;                                            /*设置左浮动*/
    width: 160px;                                           /*设置宽度*/
    height:175px;                                           /*设置高度*/
    margin: 6px;                                            /*设置外边距*/
    border: 1px solid #666666;                              /*设置边框*/
    padding-bottom: 10px;                                   /*设置下内边距*/
}
</style>
```

（3）在创建的应用程序实例中，我们定义数据和方法。其中，product 属性用于表示检索前的原产品列表。search()方法用于判断检索关键字是否为空。如果检索关键字为空，search()方法会将检索结果设置为原产品列表；否则，该方法会根据检索关键字对原产品列表 product 进行过滤，并将检索到的结果保存在 searchResult 列表中。代码如下：

```html
<script type="text/javascript">
  const vm = Vue.createApp({
    data(){
      return {
        product: [                                                          //产品列表
            {url: 'images/products/OPPO Find X7.png', name: 'OPPO Find X7'},
            {url: 'images/products/OPPO Reno11.png', name: 'OPPO Reno11'},
```

```
            {url: 'images/products/戴尔灵越 3530.jpg', name: '戴尔灵越 3530'},
            {url: 'images/products/vivo X100.png', name: 'vivo X100'},
            {url: 'images/products/华为 Mate60.png', name: '华为 Mate60'},
            {url: 'images/products/华为 nova12.png', name: '华为 nova12'},
            {url: 'images/products/华硕灵耀 14.jpg', name: '华硕灵耀 14'},
            {url: 'images/products/荣耀 100.png', name: '荣耀 100'},
            {url: 'images/products/华为擎云 S520.jpg', name: '华为擎云 S520'},
            {url: 'images/products/小米 14.png', name: '小米 14'}
        ],
        keyword: '',                                              //检索关键字
        searchResult: [],                                         //搜索结果列表
    }
},
methods: {
    search: function (){
        if(this.keyword === ''){                                  //如果检索关键字为空
            this.searchResult = this.product;                     //检索结果为原产品列表
        }else{
            var t = this;
            t.searchResult = t.product.filter(function (item){    //过滤产品列表
                if(item.name.toLowerCase().indexOf(t.keyword.toLowerCase()) !== -1){
                    return item;
                }
            });
        }
    }
},
mounted: function (){
    this.searchResult = this.product;                             //检索结果为原产品列表
}
})
</script>
```

1.4.5 浮动窗口设计

在界面右侧有一个浮动窗口，无论用户拖动界面中的横向滚动条还是纵向滚动条，该窗口的位置都保持不变。该浮动窗口主要展示"至诚服务""在线客服"和"附近门店"选项。浮动窗口效果如图1.10所示。

浮动窗口的实现步骤如下。

（1）编写HTML代码，定义<div>元素，并对该元素的style属性进行样式绑定。在该元素中，添加组成浮动窗口的多个元素。代码如下：

```
<div class="service" :style="{right: rightDis + 'px', top: topDis + 'px'}">
    <img src="images/ra_01.png">
    <div>
        <img src="images/ra_04.png">
        <span>至诚服务</span>
        <img src="images/ra_05.png">
        <span>在线客服</span>
        <img src="images/ra_06.png">
        <span>附近门店</span>
    </div>
    <img src="images/ra_02.png">
</div>
```

图1.10 浮动窗口

（2）编写CSS代码，为div元素和span元素设置样式。关键代码如下：

```
<style>
```

```css
.service{
    height:45px;                                    /*设置高度*/
    position:absolute;                              /*设置元素绝对定位*/
    width:81px;                                     /*设置宽度*/
}
.service div{
    height: 256px;                                  /*设置高度*/
    text-align: center;                             /*设置文本水平居中显示*/
    background-image: url("../images/ra_03.gif");   /*设置背景图像*/
}
.service span{
    display: inline-block;                          /*设置元素为行内块元素*/
    margin: 2px auto 10px auto;                     /*设置外边距*/
}
</style>
```

（3）在创建的应用程序实例中，我们定义数据。其中，rightDis 属性表示浮动窗口到页面右侧的距离，而 topDis 属性则表示浮动窗口到页面顶部的距离。在 mounted()钩子函数中，我们添加窗口的 onscroll 事件监听器，以便在滚动条被拖动时动态地设置浮动窗口到页面顶部和右侧的距离。代码如下：

```html
<script type="text/javascript">
    const vm = Vue.createApp({
        data(){
            return {
                rightDis: 20,                       //浮动窗口到页面右侧距离
                topDis: 100                         //浮动窗口到页面顶部距离
            }
        },
        mounted: function (){
            var t = this;
            window.onscroll = function (){
                //设置浮动窗口在垂直方向的绝对位置
                t.topDis = document.documentElement.scrollTop + 100;
                //设置浮动窗口在水平方向的绝对位置
                t.rightDis = 20 - document.documentElement.scrollLeft;
            }
        }
    })
</script>
```

> **说明**
> 在设计浮动窗口时，在 mounted()钩子函数中，我们为 Window 对象添加了 onscroll 事件监听器，由于事件处理程序中的 this 和 Vue 实例中的 this 指向不同的作用域，因此在使用 onscroll 事件监听器之前，我们需要对 this 进行重新赋值。

1.5 项目运行

通过前述步骤，我们已经设计并完成了"智汇企业官网首页设计"项目的开发。接下来，我们将运行该项目以检验我们的开发成果。我们在浏览器中打开项目文件夹中的 index.html 文件，即可成功运行该项目。运行效果如图 1.11 所示。

图1.11 智汇企业官网首页设计

1.6 源码下载

本章虽然详细地讲解了如何编码实现智汇企业官网首页的各个功能，但给出的代码都是代码片段，而非完整的源代码。为了方便读者学习，本书提供了该项目的完整源代码，读者可以通过扫描右侧的二维码进行下载。

源码下载

第 2 章 贪吃蛇小游戏

——v-show 指令 + 事件处理 + 表单元素绑定

贪吃蛇游戏是一款十分经典的大众游戏，它因操作简单、娱乐性强而受到广大游戏爱好者的欢迎。目前该游戏在计算机端和手机端都有具体应用，且新版本、变种众多。本章将使用 Vue.js 中的 v-show 指令、事件处理和表单元素绑定等技术实现一个网页版的贪吃蛇小游戏。

项目微视频

本项目的核心功能及实现技术如下：

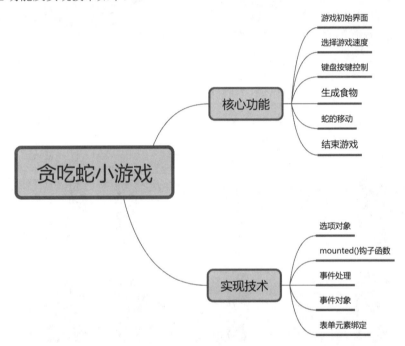

2.1 开发背景

贪吃蛇游戏是一款经典的电子游戏，诞生于 20 世纪 70 年代末到 80 年代初。它最初出现在贪吃蛇游戏机上，后来被移植到个人计算机和其他游戏平台上，成为了广受欢迎的游戏之一。贪吃蛇游戏的玩法非常简单，玩家只需要通过控制蛇的方向，使其在地图上移动并吞噬食物。随着蛇身长度的增长，玩家的分数也会增加。当蛇碰到自身或地图边界时，游戏结束。

贪吃蛇游戏早期多为单人游戏，但随着技术的发展，出现了支持多人游戏的版本。这使得玩家可以与朋友或其他玩家一同游戏，增加了社交互动和竞争的乐趣。贪吃蛇游戏不仅在过去深受欢迎，如今也仍然有许多新版本和变种出现，包括 3D 效果、特殊道具等。该游戏因其规则简单且易上手，而成为电子游戏界的经典之作，其魅力经久不衰。本章将使用 Vue.js 开发一个网页版的贪吃蛇小游戏，该项目将包含贪吃蛇

游戏最基本的页面结构和功能,其实现目标如下:
- ☑ 生成游戏地图。
- ☑ 单击空格键开始或暂停游戏。
- ☑ 通过选择游戏速度增加或降低游戏难度。
- ☑ 使用键盘中的方向键控制蛇的前进方向。
- ☑ 当蛇头撞到边界或自身时,游戏结束。

2.2 系统设计

2.2.1 开发环境

本项目的开发及运行环境如下:
- ☑ 操作系统:推荐 Windows 10、Windows 11 或更高版本,同时兼容 Windows 7(SP1)。
- ☑ 开发工具:WebStorm。
- ☑ 开发框架:Vue.js 3.0。

2.2.2 业务流程

在设计贪吃蛇小游戏时,首先需要初始化游戏界面,创建游戏地图,同时在地图中随机生成食物。在游戏开始前,玩家可以选择游戏速度,以调整游戏难度。然后,玩家通过单击空格键控制蛇开始移动或停止移动。蛇开始移动后,玩家可以通过键盘上的方向键控制蛇的移动方向。在移动的过程中,蛇头每碰到一个食物就会将其吃掉,每吃掉一个食物,蛇身会加长,玩家分数也会增加 1 分。如果蛇头碰到自身或边界,则游戏结束。

本项目的业务流程如图 2.1 所示。

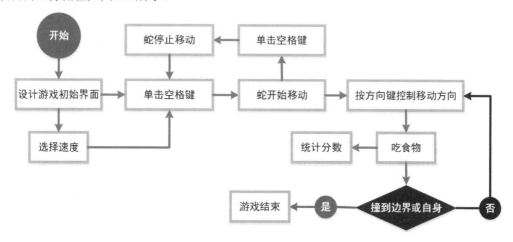

图 2.1 本项目的业务流程图

2.2.3 功能结构

本项目的功能结构已经在章首页中给出。其实现的具体功能如下:

- ☑ 初始化游戏地图：显示可移动区域，定位蛇身位置，并在地图上随机生成食物。
- ☑ 选择游戏速度：通过下拉菜单选择游戏速度，以调整游戏难度。
- ☑ 开始游戏：按空格键开始游戏，通过方向键控制蛇的移动方向。每吃掉一个食物，蛇身长度会加长，玩家分数相应增加 1 分。
- ☑ 游戏结束：当蛇头碰到自身或边界时，显示游戏结束。

2.3 技术准备

在开发贪吃蛇小游戏时，我们主要使用了 Vue.js 框架。基于此，我们将详细阐述本项目所用的 Vue.js 核心技术点及其具体作用：

- ☑ 选项对象：选项对象中包括数据、方法等选项。在开发该项目时，我们将需要应用的数据定义在 data 选项中，将实现某个功能的方法定义在 methods 选项中。
- ☑ mounted 钩子函数：这个钩子函数在 DOM 文档渲染完毕之后被调用，相当于 JavaScript 中的 window.onload()方法。在开发该项目时，我们将初始化游戏界面的方法和按下键盘按键时调用的方法放在 mounted 钩子函数中，这样在文档渲染完毕后，就可以显示初始化的游戏界面。关键代码如下：

```
<script src="https://unpkg.com/vue@3"></script>
<script type="text/javascript">
    const vm = Vue.createApp({
        mounted: function (){
                this.dom = document.getElementById("map");
                this.init();                              //调用初始化方法
                var t = this;
                document.onkeydown = function(){
                        t.trigger();                      //按下按键时调用 trigger 方法
                }
        }
    }).mount('#app');
</script>
```

- ☑ v-show 指令：v-show 指令是根据表达式的值来判断是否显示或隐藏 DOM 元素的。当表达式的值为 true 时，元素将被显示；当表达式的值为 false 时，元素将被隐藏。使用 v-show 指令的元素，无论表达式的值为 true 还是 false，该元素都始终会被渲染并保留在 DOM 中。绑定值的改变只是简单地切换元素的 CSS 属性 display。在开发该项目时，我们对用于显示游戏提示文字的<div>元素使用 v-show 指令，并通过该指令的绑定值来控制是否显示相应文字。关键代码如下：

```
<div class="show" v-show="isShow">{{result}}</div>
<script type="text/javascript">
    const vm = Vue.createApp({
        data(){
                return {
                        isShow: false,                    //是否显示游戏提示
                        result: ''                        //游戏提示信息
                }
        }
    }).mount('#app');
</script>
```

- ☑ 事件处理：在 Vue.js 中，我们使用 v-on 指令对 DOM 事件进行监听。v-on 指令后面可以跟随任何原生事件名称。该项目中，当用户单击"选择速度"下拉菜单时，会触发 change 事件，随后会调

用 setSpeed()方法。该方法负责对游戏界面进行初始化操作,实现设置游戏速度的功能。
- ☑ 表单元素绑定:在 Vue.js 中,对表单元素进行双向数据绑定需要使用 v-model 指令。在修改表单元素的值发生变化时,Vue 实例中对应的属性值也会同步更新。在该项目中,我们使用 v-model 指令将下拉菜单和表示游戏速度的 speed 属性进行绑定,确保当用户在下拉菜单中选择不同的速度选项时,绑定的属性值也会相应地更新。

《Vue.js 从入门到精通》一书详细地讲解了有关选项对象、mounted 钩子函数、v-show 指令、事件处理、表单元素绑定等基础知识。对这些知识不太熟悉的读者,可以参考该书相关章节进行学习。

2.4 游戏初始界面设计

2.4.1 创建主页

设计贪吃蛇小游戏,首先需要创建游戏的主页,并在主页中定义构成游戏界面的各个元素。创建主页的关键步骤如下。

(1)新建 index.html 文件,在该文件中编写 HTML 代码,定义一个 id 属性值为 app 的<div>元素。在这个<div>元素内部,定义一个 class 属性值为 box 的<div>元素、一个 id 属性值为 map 的表格和一个 class 属性值为 show 的<div>元素。其中:第一个<div>元素用于显示游戏分数、设置游戏速度,以及显示开始或暂停游戏的提示文字,将游戏分数和定义的 foodNum 属性进行绑定,当触发选择游戏速度的下拉菜单的 change 事件时调用 setSpeed()方法,将下拉菜单的值和定义的 speed 属性进行绑定;第二个<div>元素用于显示游戏提示信息,将游戏提示信息绑定到 result 属性。代码如下:

```html
<!DOCTYPE html>
<html lang="en">
<head>
    <meta charset="UTF-8">
<title>贪吃蛇小游戏</title>
</head>
<body>
<div id="app">
    <div class="box">
        <div class="top">
            <span>分数:{{foodNum}}</span>
            <span>
                <label for="setSpeed">选择速度:</label>
                <select id="setSpeed" v-model="speed" v-on:change="setSpeed">
                    <option value="200">慢速</option>
                    <option value="100">中速</option>
                    <option value="50">快速</option>
                </select>
            </span>
        </div>
        <div class="tip">提示:单击空格键(space)开始或暂停游戏</div>
    </div>
    <table id="map"></table>
    <div class="show" v-show="isShow">{{result}}</div>
</div>
</body>
</html>
```

(2)新建一个名为 css 的文件夹,并在该文件夹中创建一个名为 snake.css 的文件,在该文件中编写游戏界面的样式。snake.css 文件的代码如下：

```css
*{
    margin:0;                               /*设置网页外边距*/
    padding:0;                              /*设置网页内边距*/
    font-size:12px;                         /*设置文字大小*/
}
table#map {
    width:auto;                             /*设置宽度*/
    height:auto;                            /*设置高度*/
    margin:0 auto;                          /*设置外边距*/
    border-collapse:collapse;               /*设置表格的边框*/
    border-spacing:0;                       /*设置边框的间距*/
    clear:both;                             /*清除浮动*/
    background:#74AFE0;                     /*设置背景颜色*/
}
td{
    width:10px;                             /*设置宽度*/
    height:10px;                            /*设置高度*/
    border:1px solid black;                 /*设置单元格边框*/
    border-radius: 5px;                     /*设置圆角边框*/
}
.snakehead{
    background-color: orangered;            /*设置背景颜色*/
}
.snakebody{
    background-color:#FFCC00;               /*设置背景颜色*/
}
.snakefood{
    background-color: orangered;            /*设置背景颜色*/
}
.box{
    width:320px;                            /*设置宽度*/
    margin:0 auto;                          /*设置外边距*/
    padding:1em 0;                          /*设置内边距*/
    text-align: center;                     /*设置文字水平居中显示*/
}
.top span{
    height:30px;                            /*设置高度*/
    margin-right:1.5em;                     /*设置右外边距*/
    line-height:30px;                       /*设置行高*/
}
.box .tip{
    color: #F00;                            /*设置文字颜色*/
}
.show{
    background:#FFFF66;                     /*设置背景颜色*/
    width:300px;                            /*设置宽度*/
    height:40px;                            /*设置高度*/
    line-height:40px;                       /*设置行高*/
    font-weight:bolder;                     /*设置字体粗细*/
    border:1px solid #999999;               /*设置边框*/
    margin:10px auto;                       /*设置外边距*/
    font-size: 20px;                        /*设置文字大小*/
    text-align: center;                     /*设置文字水平居中显示*/
}
```

（3）在 index.html 文件中首先引入 snake.css 文件，然后使用 CDN 方式引入 Vue.js。代码如下：

```html
<link rel="stylesheet" href="css/snake.css">
<script src="https://unpkg.com/vue@3"></script>
```

2.4.2 游戏初始化

在游戏初始界面,贪吃蛇的可移动区域是一个 21 行 21 列的表格。表格上方显示的内容包括游戏初始分数、用于选择游戏速度的下拉菜单,以及开始或暂停游戏的提示文字。游戏初始界面的效果如图 2.2 所示。

对游戏进行初始化的实现步骤如下。

(1) 在 index.html 文件中编写 JavaScript 代码。创建应用程序实例,定义数据和方法。首先在 data 选项中对多个属性进行初始化。然后在 methods 选项中定义用于创建贪吃蛇可移动区域的 map() 方法,在该方法中应用 Table 对象的 insertRow() 方法和 insertCell() 方法创建表格行和单元格。接下来定义用于生成食物的 food() 方法。最后定义 init() 方法,在该方法中调用 map() 方法和 food() 方法实现游戏的初始化。代码如下:

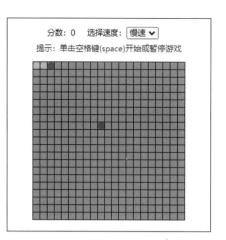

图 2.2　游戏初始界面效果

```javascript
<script type="text/javascript">
    const vm = Vue.createApp({
        data(){
            return {
                rows: 21,                          //21 行
                cols: 21,                          //21 列
                speed: 200,                        //前进速度
                curKey: 0,                         //当前方向按键键码值
                timer: 0,
                pos: [],                           //蛇身位置
                foodPos: {"x":-1,"y":-1},
                foodNum: -1,                       //吃掉食物数量
                dom: null,                         //地图元素
                pause: 1,                          //1 表示暂停,-1 表示开始
                isShow: false,                     //是否显示游戏提示
                result: ''                         //游戏提示信息
            }
        },
        methods: {
            map: function(){                       //创建地图
                if(this.dom.firstChild){
                    this.dom.removeChild(this.dom.firstChild);  //重新开始,删除之前创建的 tbody
                }
                for( j = 0; j < this.rows; j++ ){
                    var tr = this.dom.insertRow(-1);            //插入一行
                    for( i = 0; i < this.cols; i++ ){
                        tr.insertCell(-1);                      //插入一列
                    }
                }
            },
            food: function(){                      //生成食物
                do{
                    this.foodPos.y = Math.floor( Math.random()*this.rows );
                    this.foodPos.x = Math.floor( Math.random()*this.cols );
                }while( this.dom.rows[this.foodPos.y].cells[this.foodPos.x].className !== "" )  //防止食物生成在蛇身上
                this.dom.rows[this.foodPos.y].cells[this.foodPos.x].className="snakefood";  //设置食物样式
                this.foodNum++;                    //设置分数
            },
            init: function(){
                this.map();                        //创建地图
```

```
                window.clearInterval(this.timer);                                        //停止
                this.pos = [{"x":2,"y":0},{"x":1,"y":0},{"x":0,"y":0}];                   //定义蛇身位置
                for(var j=0; j<this.pos.length; j++ ){                                    //显示蛇身
                    this.dom.rows[this.pos[j].y].cells[this.pos[j].x].className="snakebody";
                }
                this.dom.rows[this.pos[0].y].cells[this.pos[0].x].className="snakehead";  //为蛇头设置样式
                this.curKey = 0;                                                          //当前方向按键键码值
                this.foodNum = -1;                                                        //吃掉食物数量
                this.food();                                                              //生成食物
                this.pause = 1;                                                           //1：暂停；-1：开始
            }
        }
    }).mount('#app');
</script>
```

（2）在 mounted 钩子函数中获取贪吃蛇的移动区域，并调用 init()方法对游戏进行初始化。代码如下：

```
mounted: function (){
    this.dom = document.getElementById("map");
    this.init();                                                                          //调用初始化方法
}
```

说明

在该项目中，对游戏进行初始化时，我们使用了 JavaScript 中的 Table 对象。其中，insertRow()方法用于在表格中插入一个新行，insertCell()方法用于向行中插入一个单元格，rows[]集合用于返回包含表格中所有行的一个数组，cells[]集合用于返回包含表格中所有单元格的一个数组。

2.4.3　设置游戏速度

在游戏初始界面，玩家可以通过贪吃蛇可移动区域上方的下拉菜单来设置游戏速度。该游戏提供了 3 种可选的游戏速度：慢速、中速和快速。在 Vue 实例的 methods 选项中，我们定义 setSpeed()方法，当用户选择下拉菜单中的不同速度选项并触发 change 事件时，该方法会被调用，执行相应的事件处理逻辑，代码如下：

```
setSpeed: function (event){
    event.target.blur();                                                                  //使下拉菜单失去焦点
    this.init();                                                                          //调用初始化方法
    this.isShow = false;                                                                  //隐藏游戏提示
}
```

2.5　游戏操作

在对游戏进行初始化后，玩家只需单击空格键，贪吃蛇便会开始移动。通过键盘中的方向键，玩家可以控制贪吃蛇的移动方向。在贪吃蛇的移动过程中，玩家再次单击空格键可以暂停贪吃蛇的移动，而再次单击空格键则可以使贪吃蛇继续移动。

2.5.1　键盘按键控制

在 methods 选项中，我们定义 trigger()方法，该方法用于控制贪吃蛇的移动方向，并控制游戏的开始和

暂停。代码如下：

```
trigger: function(){
    var _t=this;
    var eKey = event.keyCode;                                            //获取按键键码值
    //如果按下的是方向键，并且不是当前方向，也不是反方向和暂停状态
    if( eKey>=37 && eKey<=40 && eKey!==this.curKey && !( (this.curKey === 37 && eKey === 39)
        || (this.curKey === 38 && eKey === 40) || (this.curKey === 39 && eKey === 37) ||
        (this.curKey === 40 && eKey === 38) ) && this.pause===-1 ){
        this.curKey = eKey;                                              //设置当前方向按键键码值
    }else if( eKey===32 ){
        this.curKey = (this.curKey===0) ? 39 : this.curKey;
        this.pause*=-1;
        if(this.pause===-1){
            this.timer=window.setInterval(function(){_t.move()},this.speed);  //设置蛇身移动
            this.isShow = true;                                          //显示游戏提示
            this.result = "游戏进行中";
        }else{
            window.clearInterval(this.timer);                            //停止蛇的移动
            this.result = "游戏暂停";
        }
    }
}
```

2.5.2 蛇的移动

在 methods 选项中，我们定义 move()方法，该方法用于实现贪吃蛇的移动、贪吃蛇吃食物，以及判断蛇头是否撞到边界或自身的操作，代码如下：

```
move: function(){                                                        //移动蛇头
    switch(this.curKey){
        case 37:                                                         //左方向
            if( this.pos[0].x <= 0 ){                                    //蛇头撞到边界
                this.over();
                return;
            }else{
                this.pos.unshift( {"x":this.pos[0].x-1,"y":this.pos[0].y}); //添加元素
            }
            break;
        case 38:                                                         //上方向
            if( this.pos[0].y <= 0 ){
                this.over();
                return;
            }else{
                this.pos.unshift( {"x":this.pos[0].x,"y":this.pos[0].y-1});
            }
            break;
        case 39:                                                         //右方向
            if( this.pos[0].x >= this.cols-1 ){
                this.over();
                return;
            }else{
                this.pos.unshift( {"x":this.pos[0].x+1,"y":this.pos[0].y});
            }
            break;
        case 40:                                                         //下方向
            if( this.pos[0].y >= this.rows-1 ){
                this.over();
                return;
```

```
            }else{
                    this.pos.unshift( {"x":this.pos[0].x,"y":this.pos[0].y+1});
            }
                    break;
    }
    if( this.pos[0].x === this.foodPos.x && this.pos[0].y === this.foodPos.y ){    //蛇头位置与食物重叠
        this.food();                                                                //生成食物
    }else if( this.curKey !== 0 ){
        this.dom.rows[this.pos[this.pos.length-1].y].cells[this.pos[this.pos.length-1].x].className="";
        this.pos.pop();                                                             //删除蛇尾
    }
    for(i=3;i<this.pos.length;i++){                                                 //从蛇身的第四节开始判断是否撞到自己
        if( this.pos[i].x === this.pos[0].x && this.pos[i].y === this.pos[0].y ){
            this.over();                                                            //游戏结束
            return;
        }
    }
    this.dom.rows[this.pos[0].y].cells[this.pos[0].x].className="snakehead";        //画新蛇头
    this.dom.rows[this.pos[1].y].cells[this.pos[1].x].className="snakebody";        //原蛇头变为蛇身
}
```

在 mounted 钩子函数中，我们设置一个事件监听器，当触发 onkeydown 事件时，调用 trigger()方法。代码如下：

```
document.onkeydown = function(){
    t.trigger();                                                                    //按下按键时调用 trigger()方法
}
```

2.5.3 游戏结束

在贪吃蛇移动时，如果蛇头撞到可移动区域的边界或自己的身体，则游戏结束。在 methods 选项中，我们定义一个名为 over()的方法来处理游戏结束时的情况。游戏结束后，玩家将有 3 秒的时间来准备重新开始游戏。代码如下：

```
over: function(){
    this.result = "游戏结束";
    window.clearInterval(this.timer);                                               //停止蛇的移动
    document.onkeydown = null;
    var t = this;
    setTimeout(function (){
        t.isShow = false;                                                           //隐藏游戏提示
        t.init();                                                                   //重置游戏状态
        document.onkeydown = function(){
            t.trigger();                                                            //按下按键时调用 trigger()方法
        }
    },3000);
}
```

2.6 项目运行

通过前述步骤，我们已经设计并完成了"贪吃蛇小游戏"项目的开发。接下来，我们运行该项目，以检验我们的开发成果。在浏览器中打开项目文件夹中的 index.html 文件，即可成功运行该项目。游戏进行中的界面效果如图 2.3 所示。游戏暂停时的界面效果如图 2.4 所示。游戏结束时的界面效果如图 2.5 所示。

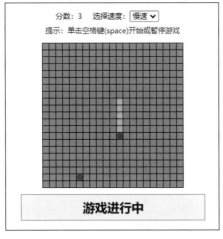
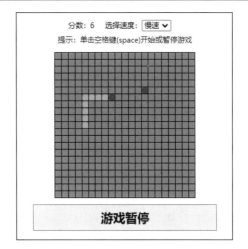

图 2.3　游戏进行中的界面效果　　　　图 2.4　游戏暂停时的界面效果

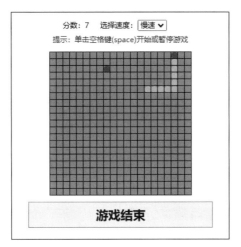

图 2.5　游戏结束时的界面效果

2.7　源码下载

本章虽然详细地讲解了如何编码实现"贪吃蛇小游戏"的各个功能，但给出的代码都是代码片段，而非完整的源代码。为了方便读者学习，本书提供了该项目的完整源代码，读者可以扫描右侧的二维码进行下载。

源码下载

第 3 章 时光音乐网首页设计

——Vue CLI + axios

音乐类网站是当今流行音乐的主流发布平台。随着网络技术的发展，用户对音乐网站的要求逐渐提高。音乐网站为了吸引更多的用户，需要对界面进行不断的美化。本章将使用 Vue CLI 和 axios 设计一个简洁大方的音乐网首页。

项目微视频

本项目的核心功能及实现技术如下：

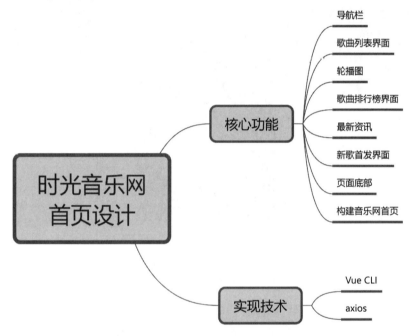

3.1 开发背景

在人们的日常生活中，音乐已经成为不可或缺的一部分。随着互联网技术的不断发展，人们对于音乐的需求也日益增加，这促使在线音乐网站应运而生。该类网站可以为用户提供丰富的音乐资源和个性化的音乐推荐，使人们能在线欣赏不同风格、不同国家的歌曲，实现资源共享。本章将使用 Vue CLI 和 axios 开发一个音乐网站的首页，其实现目标如下：

- ☑ 通过歌曲列表展示歌曲信息。
- ☑ 通过轮播图展示歌曲封面。
- ☑ 根据选项卡切换不同类别的歌曲列表。
- ☑ 间断滚动最新音乐资讯。
- ☑ 分页展示首发歌曲信息。

3.2 系统设计

3.2.1 开发环境

本项目的开发及运行环境如下：
- ☑ 操作系统：推荐 Windows 10、Windows 11 或更高版本，同时兼容 Windows 7（SP1）。
- ☑ 开发工具：WebStorm。
- ☑ 开发框架：Vue.js 3.0。

3.2.2 业务流程

时光音乐网首页由多个单文件组件构成，包括导航栏组件、歌曲列表组件、轮播图组件、歌曲排行榜组件、最新音乐资讯组件、新歌首发组件和页面底部组件。

根据该网站首页的业务需求，我们设计如图 3.1 所示的业务流程图。

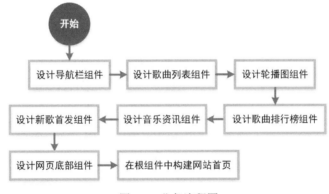

图 3.1 业务流程图

3.2.3 功能结构

本项目的功能结构已经在章首页中给出，其实现的具体功能如下：
- ☑ 导航栏：鼠标指向某个导航菜单项，下方显示该菜单项对应的子项。
- ☑ 歌曲列表：展示歌曲信息，包括歌曲图片、歌曲名称和歌曲简介。
- ☑ 轮播图：单击某个数字按钮可切换到对应的图片。
- ☑ 歌曲排行榜：根据选项卡切换不同类别的歌曲列表。
- ☑ 最新音乐资讯：从下到上间断循环滚动音乐资讯。
- ☑ 新歌首发：分页展示首发歌曲信息。
- ☑ 页面底部：展示版权等方面的信息。

3.3 技术准备

在开发时光音乐网首页时，我们主要应用 Vue CLI 脚手架工具和 axios。下面将简要概述本项目所用的

Vue.js 核心技术点及其各自的具体作用。

1. 单文件组件

将一个组件的 HTML、JavaScript 和 CSS 应用各自的标签写在一个文件中，这样的文件即为单文件组件。单文件组件是 Vue 自定义的一种文件，以.vue 作为文件的扩展名。示例代码如下：

```
<template>
    <p>{{ msg }}</p>
</template>
<script>
    export default {
        data: function () {
            return {
                msg: '长风破浪会有时，直挂云帆济沧海。'
            }
        }
    }
</script>
<style scoped>
    p {
        font-size: 36px;                    /*设置文字大小*/
        text-align: center;                 /*设置文字居中显示*/
        color: #FF00FF;                     /*设置文字颜色*/
    }
</style>
```

2. Vue CLI

Vue CLI 是一个基于 Vue.js 进行快速开发的完整系统。最新版本的 Vue CLI 的包名从原来的 vue-cli 被更改为了@vue/cli。Vue CLI 是使用 Node.js 编写的命令行工具，需要进行全局安装来使用。要安装最新版本的 Vue CLI，需要在命令提示符窗口中输入如下命令：

```
npm install -g @vue/cli
```

要使用 Vue CLI 创建项目，可以使用 vue create 命令。例如，要创建一个名称为 myapp 的项目，需要输入以下命令：

```
vue create myapp
```

执行该命令后只需遵循命令行工具提供的指示步骤，就能顺利完成项目的创建。

3. axios

axios 是一个基于 Promise 的 HTTP 客户端，它既可以在浏览器中使用，也可以在 Node.js 中运行。使用 axios，可以实现 Ajax 请求，进而实现本地与服务器之间的通信。在项目中使用 axios 时，可以通过 npm 方式进行安装。在命令提示符窗口中，输入如下命令进行安装：

```
npm install axios --save
```

安装 axios 之后，需要在项目中引入它，代码如下：

```
import axios from 'axios'
```

axios 常用的请求方法包括 GET 和 POST 等。其中：GET 请求主要用于从服务器上获取数据，传递的数据量比较小；POST 请求主要用于向服务器传递数据，传递的数据量比较大。使用 axios 无论发送 GET 请求还是 POST 请求，在发送请求后都需要使用回调函数对请求的结果进行处理。如果请求成功，则需要使用.then()方法处理请求的结果；如果请求失败，则需要使用.catch()方法处理请求的结果。示例代码如下：

```
axios.get('/book',{
    params:{                              //传递的参数
```

```
            type : 'Vue',
            number : 10
        }
}).then(function(response){
    console.log(response.data);
}).catch(function(error){
    console.log(error);
})
```

> **注意**
> 这两个回调函数各自拥有独立的作用域，如果在它们内部访问 Vue 实例，就不能直接使用 this 关键字。为了解决这个问题，需要在回调函数的后面添加.bind(this)来确保 this 指向正确的 Vue 实例。

有关《Vue.js 从入门到精通》一书详细地讲解了单文件组件、Vue CLI、axios 等基础知识。对于这些知识不太熟悉的读者，可以参考该书的相应章节进行深入学习。

3.4 功能设计

设计时光音乐网首页各功能模块之前，需要先使用 Vue CLI 创建项目，项目名称设置为 music。创建项目之后，即可开始首页中各功能模块的设计。

3.4.1 导航栏的设计

网站首页的导航栏包含 6 个菜单项。除了"首页"，还有 5 个菜单项，分别表示 5 种不同类型的音乐风格：流行音乐、摇滚音乐、民族音乐、古典音乐和金属乐。当鼠标悬停在某个导航菜单项上时，其下方会显示对应的子项。网站首页的导航栏效果如图 3.2 所示。

首页	流行音乐	摇滚音乐	民族音乐	古典音乐	金属乐
		硬核摇滚 艺术摇滚 朋克			

图 3.2 网站首页的导航栏效果

设计导航栏的关键步骤如下。

（1）在 components 目录中创建 nav 文件夹，在该文件夹下创建 6 个与导航菜单项对应的子项组件。然后在 components 目录中创建 MusicNav.vue 文件。在 MusicNav.vue 文件的<template>标签中定义导航栏，将导航菜单项的类名和定义的数据进行绑定。当鼠标悬停在菜单项上时，触发 mouseover 事件，并通过为数据赋值的方式改变当前渲染的组件，再使用动态组件的形式实现 6 个子项组件之间的切换。在<script>标签中引入 6 个子项组件，并对 6 个组件进行注册。具体代码如下：

```
<template>
  <div class="menu">
    <div class="main_menu">
      <ul>
        <li :class="{active: current === 'Home'}" @mouseover="current = 'Home'">首页</li>
        <li :class="{active: current === 'Pop'}" @mouseover="current = 'Pop'">流行音乐</li>
        <li :class="{active: current === 'Rock'}" @mouseover="current = 'Rock'">摇滚音乐</li>
        <li :class="{active: current === 'Nation'}" @mouseover="current = 'Nation'">民族音乐</li>
        <li :class="{active: current === 'Classical'}" @mouseover="current = 'Classical'">古典音乐</li>
        <li :class="{active: current === 'Metal'}" @mouseover="current = 'Metal'">金属乐</li>
      </ul>
```

```
        </div>
        <div class="sub_menu">
            <component :is="current"></component>
        </div>
    </div>
</template>
<script>
import Home from "@/components/nav/HomeNav";
import Pop from "@/components/nav/PopNav";
import Rock from "@/components/nav/RockNav";
import Nation from "@/components/nav/NationNav";
import Classical from "@/components/nav/ClassicalNav";
import Metal from "@/components/nav/MetalNav";
export default {
    name: "MusicNav",
    data: function (){
        return {
            current: 'Home'                          //当前导航菜单项组件名
        }
    },
    components: {
        Home,                                        //注册 Home 组件
        Pop,                                         //注册 Pop 组件
        Rock,                                        //注册 Rock 组件
        Nation,                                      //注册 Nation 组件
        Classical,                                   //注册 Classical 组件
        Metal                                        //注册 Metal 组件
    }
}
</script>
```

（2）在 assets 目录中创建 css 文件夹，在该文件夹下创建 style.css 文件，在该文件中编写 CSS 代码，为导航栏组件设置样式。代码如下：

```css
.main_menu{
    width: 1000px;                                   /*设置宽度*/
    margin: 0 auto;                                  /*设置外边距*/
    border-top:solid 1px #D0D0D0;                    /*设置上边框*/
    border-bottom:solid 1px #D0D0D0;                 /*设置下边框*/
    font-weight:bold;                                /*设置文字粗细*/
    letter-spacing:2px;                              /*设置文字间距*/
}
ul,li{
    list-style: none;                                /*设置列表无样式*/
}
.main_menu li{
    float: left;                                     /*设置左浮动*/
    width: 165px;                                    /*设置宽度*/
    height: 42px;                                    /*设置高度*/
    line-height: 42px;                               /*设置行高*/
    text-align: center;                              /*设置文本水平居中显示*/
    cursor: pointer;                                 /*设置鼠标形状*/
}
.main_menu li:not(:last-child){
    border-right: 1px solid #D0D0D0;                 /*设置右边框*/
}
.sub_menu{
    width: 1000px;                                   /*设置宽度*/
    margin: 0 auto;                                  /*设置外边距*/
    height: 42px;                                    /*设置高度*/
}
.sub_menu span,.sub_menu li{
```

```
        height: 42px;                       /*设置高度*/
        line-height: 42px;                  /*设置行高*/
}
.sub_menu li{
        float: left;                        /*设置左浮动*/
        text-align:left;                    /*设置文本水平向左显示*/
        padding-left:12px;                  /*设置左内边距*/
        margin-left:25px;                   /*设置左外边距*/
}
```

3.4.2 歌曲列表展示界面

在时光音乐网首页中，主显示区域的左侧是一个歌曲列表展示界面。该界面共展示了 9 首歌曲，每首歌曲都配有对应的歌曲图片、歌曲名称和歌曲简介。界面效果如图 3.3 所示。

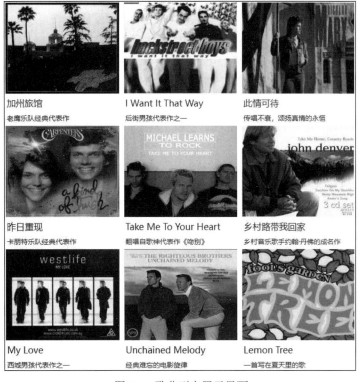

图 3.3 歌曲列表展示界面

歌曲列表展示界面的实现步骤如下。

（1）在 components 目录中创建 SongList.vue 文件。在 SongList.vue 文件的<template>标签中，定义一个 ul 列表，并使用 v-for 指令对保存歌曲信息的数组 song_list 进行遍历。在<script>标签中，定义一个歌曲列表数组 song_list，该数组包括歌曲图片 URL、歌曲名称和歌曲简介等相关信息。具体代码如下：

```
<template>
    <div class="left">
        <ul v-for="value in song_list" :key="value">
            <li><img :src="value.imgUrl"></li>
            <li>{{value.name}}</li>
            <li>{{value.intro}}</li>
        </ul>
    </div>
</template>
```

```
<script>
export default {
  name: "SongList",
  data: function (){
    return {
      song_list: [                                      //歌曲列表数组
        {
          imgUrl: require('@/assets/images/1.jpg'),     //歌曲图片 URL
          name: '加州旅馆',                              //歌曲名称
          intro: '老鹰乐队经典代表作'                     //歌曲简介
        },
        {
          imgUrl: require('@/assets/images/2.jpg'),
          name: 'I Want It That Way',
          intro: '后街男孩代表作之一'
        },
        {
          imgUrl: require('@/assets/images/3.jpg'),
          name: '此情可待',
          intro: '传唱不衰，颂扬真情的永恒'
        },
        {
          imgUrl: require('@/assets/images/4.jpg'),
          name: '昨日重现',
          intro: '卡朋特乐队经典代表作'
        },
        {
          imgUrl: require('@/assets/images/5.jpg'),
          name: 'Take Me To Your Heart',
          intro: '翻唱自歌神代表作《吻别》'
        },
        {
          imgUrl: require('@/assets/images/6.jpg'),
          name: '乡村路带我回家',
          intro: '乡村音乐歌手约翰·丹佛的成名作'
        },
        {
          imgUrl: require('@/assets/images/7.jpg'),
          name: 'My Love',
          intro: '西城男孩代表作之一'
        },
        {
          imgUrl: require('@/assets/images/8.jpg'),
          name: 'Unchained Melody',
          intro: '经典难忘的电影旋律'
        },
        {
          imgUrl: require('@/assets/images/9.jpg'),
          name: 'Lemon Tree',
          intro: '一首写在夏天里的歌'
        }
      ]
    }
  }
}
</script>
```

（2）在 style.css 文件中编写 CSS 代码，为歌曲列表组件设置样式。代码如下：

```
.main .left{
    float: left;                    /*设置左浮动*/
    width: 720px;                   /*设置宽度*/
}
.main .left img{
```

```
        width: 230px;                           /*设置宽度*/
        height: 180px;                          /*设置高度*/
}
.main .left ul{
        float: left;                            /*设置左浮动*/
        width: 240px;                           /*设置宽度*/
}
.main .left ul li:nth-child(2){
        height: 30px;                           /*设置高度*/
        line-height: 30px;                      /*设置行高*/
        font-size: 18px;                        /*设置文字大小*/
        margin-left: 3px;                       /*设置左外边距*/
}
.main .left ul li:nth-child(3){
        height: 30px;                           /*设置高度*/
        line-height: 30px;                      /*设置行高*/
        font-size: 14px;                        /*设置文字大小*/
        margin-left: 3px;                       /*设置左外边距*/
}
```

3.4.3 轮播图的设计

在首页主显示区右侧的最上方，有一个轮播图的展示界面。该界面有 3 张轮播图片，每张轮播图片下方都有 3 个数字按钮，用户可以通过单击其中一个数字按钮来切换到对应的图片。界面效果分别如图 3.4 和图 3.5 所示。

图 3.4 轮播图 1

图 3.5 轮播图 3

轮播图的实现步骤如下。

（1）在 components 目录中创建 MusicBanner.vue 文件。在 MusicBanner.vue 文件的 <template> 标签中定义轮播图片和用于切换图片的数字按钮。在 <script> 标签中定义数据、方法和钩子函数。在方法中，通过 next() 方法设置下一张图片的索引，通过 toggle() 方法设置当单击某个数字按钮后显示对应的图片。在 mounted 钩子函数中使用 setInterval() 方法每隔 3 秒调用一次 next() 方法，以实现图片自动轮播的效果。具体代码如下：

```
<template>
  <div class="right">
    <!--切换的图片-->
    <div class="banner">
      <transition-group name="effect" tag="div">
        <span v-for="(v,i) in bannerURL" :key="i" v-show="(i+1)===index?true:false">
          <img :src="v">
        </span>
      </transition-group>
```

```html
        </div>
        <!--切换的小按钮-->
        <ul class="numBtn">
            <li v-for="num in 3" :key="num">
                <a href="javascript:;" :style="{background:num===index?'#ff9900':'#CCCCCC'}"
                    @click='toggle(num)' class='num'>{{num}}</a>
            </li>
        </ul>
    </div>
</template>

<script>
export default {
    name: "MusicBanner",
    data: function (){
        return {
            bannerURL : [
                require('@/assets/images/10.jpg'),
                require('@/assets/images/11.jpg'),
                require('@/assets/images/12.jpg')
            ],                                              //图片 URL 数组
            index : 1,                                      //图片的索引
            flag : true,                                    //是否可以单击数字按钮
            timer : null,                                   //定时器 ID
        }
    },
    methods: {
        next : function(){
            this.index = this.index + 1 === 4 ? 1 : this.index + 1;
        },
        toggle : function(num){
            //单击按钮切换到对应图片
            if(this.flag){
                this.flag = false;
                //过 1 秒后可以再次单击按钮切换图片
                setTimeout(()=>{
                    this.flag = true;
                },1000);
                this.index = num;                           //切换为选中的图片
                clearTimeout(this.timer);                   //取消定时器
                //过 3 秒图片轮换
                this.timer = setInterval(this.next,3000);
            }
        }
    },
    mounted : function(){
        this.timer = setInterval(this.next,3000);           //过 3 秒图片轮换
    }
}
</script>
```

（2）在 style.css 文件中编写 CSS 代码，为轮播图组件设置样式。代码如下：

```css
.right{
    float: right;                   /*设置右浮动*/
    position: relative;             /*设置相对定位*/
    width: 260px;                   /*设置宽度*/
}
.banner img{
    width: 260px;                   /*设置宽度*/
}
.banner{
    position: relative;             /*设置相对定位*/
```

```css
        height: 260px;                                      /*设置高度*/
}
.banner span{
        position: absolute;                                 /*设置绝对定位*/
        top:0;                                              /*设置距离父元素顶部的距离*/
        left: 0;                                            /*设置距离父元素左侧的距离*/
}
.numBtn{
        width: 200px;                                       /*设置宽度*/
        position:absolute;                                  /*设置绝对定位*/
        left:50%;                                           /*设置距离父元素左侧的距离*/
        top:86%;                                            /*设置距离父元素顶部的距离*/
        text-align:center;                                  /*设置文字居中显示*/
}
.numBtn li{
        list-style:none;                                    /*设置列表无样式*/
        border-radius: 50%;                                 /*设置圆角边框*/
        float:left;                                         /*设置左浮动*/
}
.numBtn li a{
        display: block;                                     /*设置块状元素*/
        width: 20px;                                        /*设置宽度*/
        height: 20px;                                       /*设置高度*/
        line-height: 20px;                                  /*设置行高*/
        border-radius: 50%;                                 /*设置圆角边框*/
        margin: 5px;                                        /*设置外边距*/
        color:#FFFFFF;                                      /*设置文字颜色*/
        font-weight:bolder;                                 /*设置文字粗细*/
        text-decoration:none;                               /*设置文字无下画线*/
}
.numBtn li a.num{
        transition:all .6s ease;                            /*设置按钮过渡效果*/
}
/*设置过渡属性*/
.effect-enter-active, .effect-leave-active{
        transition: all 1s;
}
.effect-enter-from, .effect-leave-to{
        opacity: 0;
}
```

3.4.4 歌曲排行榜

在首页主显示区右侧的中间是歌曲排行榜的展示界面。这里使用了选项卡切换的效果。当单击"推荐歌曲"选项卡时,该选项卡下方会显示推荐歌曲列表;当单击"热门歌曲"选项卡时,该选项卡下方会显示热门歌曲列表。界面效果分别如图 3.6 和图 3.7 所示。

歌曲排行榜展示界面的实现步骤如下。

图 3.6 推荐歌曲列表    图 3.7 热门歌曲列表

(1)在 components 目录中创建 tab 文件夹,并在该文件夹下创建两个与选项卡对应的歌曲列表组件,分别是推荐歌曲列表组件 RecommendTab.vue 和热门歌曲列表组件 HotTab.vue。

推荐歌曲列表组件的代码如下:

```
<template>
  <div class="songlist">
    <div v-for="(item,index) in recommend" :key="index">
      <span>{{++index}}</span>
      <span>{{item.name}}</span>
      <span>{{item.singer}}</span>
    </div>
  </div>
</template>
<script>
export default {
  name: "RecommendTab",
  data: function (){
    return {
      recommend : [                                              //推荐歌曲数组
        { name : '寂静之声', singer : '保罗·西蒙' },
        { name : '英雄', singer : '玛丽亚·凯莉' },
        { name : '说你、说我', singer : '莱昂纳尔·里奇' },
        { name : '卡萨布兰卡', singer : '贝蒂·希金斯' },
        { name : '加州旅馆', singer : '老鹰乐队' },
        { name : '嘿，朱迪', singer : '披头士乐队' },
        { name : '尊重', singer : '奥蒂斯·雷' },
        { name : '你若即若离', singer : 'U2 乐队' }
      ]
    }
  }
}
</script>
```

热门歌曲列表组件的代码如下：

```
<template>
  <div class="songlist">
    <div v-for="(item,index) in hot" :key="index">
      <span>{{++index}}</span>
      <span>{{item.name}}</span>
      <span>{{item.singer}}</span>
    </div>
  </div>
</template>
<script>
export default {
  name: "HotTab",
  data: function (){
    return {
      hot : [                                                    //热门歌曲数组
        { name : '我希望它保持那样', singer : '后街男孩' },
        { name : '想念你', singer : '滚石乐队' },
        { name : '快乐的结局', singer : '艾薇儿·拉维尼' },
        { name : '寂静之声', singer : '保罗·西蒙' },
        { name : '我的爱', singer : '西城男孩' },
        { name : '管不住的音符', singer : '丹尼尔·波特' },
        { name : '情歌', singer : '阿黛尔·阿德金斯' },
        { name : '远离家乡', singer : '舞动精灵乐团' }
      ]
    }
  }
}
</script>
```

（2）在 components 目录中创建 MusicTabs.vue 文件。在该文件的<template>标签中，定义"推荐歌曲"和"热门歌曲"选项卡，并应用<component>元素将 data 中的 curtab 属性动态地绑定到它的 is 属性上。在

\<script\>标签中,引入两个歌曲列表组件,并定义数据和方法,以便对这两个歌曲列表组件进行注册。具体代码如下:

```
<template>
  <div class="tabs">
    <div class="top">
      <div class="title">歌曲排行榜</div>
      <ul class="tab">
        <li :class="{actived : actived}" v-on:mouseover="toggleAction('recommend')">推荐歌曲</li>
        <li :class="{actived : !actived}" v-on:mouseover="toggleAction('hot')">热门歌曲</li>
      </ul>
    </div>
    <component :is="curtab"></component>
  </div>
</template>
<script>
import recommend from "@/components/tab/RecommendTab";
import hot from "@/components/tab/HotTab";
export default {
  name: "MusicTabs",
  data: function (){
    return {
      actived : true,                                      //是否使用标签类名
      curtab : 'recommend'                                 //歌曲排行榜组件名
    }
  },
  methods: {
    toggleAction : function(value){
      this.curtab = value;                                 //获取当前组件名
      value === 'recommend' ? this.actived = true : this.actived = false;
    }
  },
  components: {
    recommend,
    hot
  }
}
</script>
```

(3) 在 style.css 文件中编写 CSS 代码,为歌曲排行榜组件设置样式。代码如下:

```
.tabs{
    float: right;                                          /*设置右浮动*/
    width:260px;                                           /*设置宽度*/
    margin:20px auto;                                      /*设置外边距*/
}
.top{
    height:26px;                                           /*设置高度*/
    line-height: 26px;                                     /*设置行高*/
}
.title{
    display:inline-block;                                  /*设置行内块元素*/
    font-size:18px;                                        /*设置文字大小*/
}
ul.tab{
    display:inline-block;                                  /*设置行内块元素*/
    list-style:none;                                       /*设置列表样式*/
    margin-left:20px;                                      /*设置左外边距*/
}
ul.tab li{
    margin: 0;                                             /*设置外边距*/
    padding: 0;                                            /*设置内边距*/
    float:left;                                            /*设置左浮动*/
```

```css
        width:80px;                              /*设置宽度*/
        height: 26px;                            /*设置高度*/
        line-height: 26px;                       /*设置行高*/
        font-size:14px;                          /*设置文字大小*/
        cursor:pointer;                          /*设置鼠标光标形状*/
        text-align:center;                       /*设置文本居中显示*/
    }
    ul.tab li.actived{
        display:block;                           /*设置块元素*/
        width:60px;                              /*设置宽度*/
        height: 26px;                            /*设置高度*/
        line-height: 26px;                       /*设置行高*/
        background-color:#66CCFF;                /*设置背景颜色*/
        color:#FFFFFF;                           /*设置文字颜色*/
        cursor:pointer;                          /*设置鼠标光标形状*/
    }
    .songlist{
        clear:both;                              /*设置清除浮动*/
        margin-top:10px;                         /*设置上外边距*/
    }
    .songlist div{
        width:260px;                             /*设置宽度*/
        height:31px;                             /*设置高度*/
        line-height:31px;                        /*设置行高*/
        border-bottom-width: 1px;                /*设置下边框宽度*/
        border-bottom-style: dashed;             /*设置下边框样式*/
        border-bottom-color: #333333;            /*设置下边框颜色*/
        background-color: #FFFFFF;               /*设置背景颜色*/
        font-size:14px;                          /*设置文字大小*/
    }
    .songlist div span{
        margin-left:10px;                        /*设置左外边距*/
    }
    .songlist div span:last-child{
        float:right;                             /*设置右浮动*/
        margin-right:10px;                       /*设置右外边距*/
    }
```

3.4.5 最新音乐资讯

在首页主显示区右侧的最下方是最新音乐资讯的展示界面，其中包含 5 条音乐资讯，这些资讯从下到上间断地循环滚动。为了展示这种滚动效果，该界面一次只展示 4 条音乐资讯。具体的界面效果如图 3.8 所示。

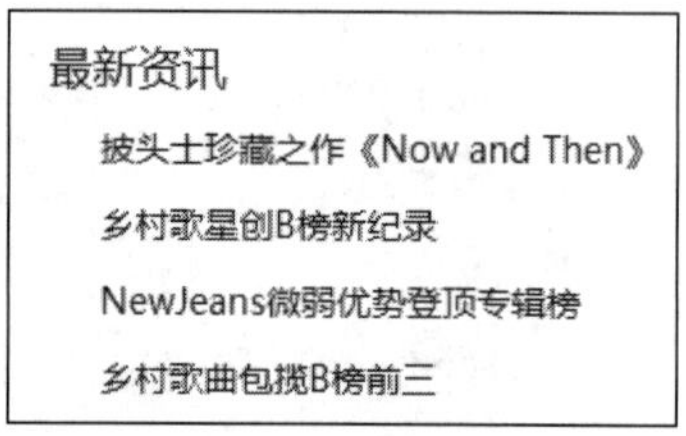

最新音乐资讯的展示界面使用 axios 发送请求和获取响应数据。因此，首先需要安装并引入 axios。实现最新音乐资讯的展示界面步骤如下：

（1）在 components 目录中创建 MusicNews.vue 文件。在该文件的 <template> 标签中定义一个 ul 列表，并将该列表的类名和定义的数据进行绑定。当鼠标移入列表时，触发 mouseenter 事件并调用 stop() 方法；当鼠标移出列表时，触发 mouseleave 事件并调用 up() 方法。在列表项中，使用 v-for 指令对音乐资讯列表 news_list 进行遍历。在 <script> 标签中定义数据、方法和钩子函数，在钩子函数中，使用 axios 发送 GET 请求，以获取 data.json 中的音乐资讯列表数据。代码如下：

图 3.8 最新音乐资讯的展示界面

```
<template>
  <div class="news">
    <div class="news_title">最新资讯</div>
    <div class="scroll">
      <ul class="list" :class="{anim:animate}" @mouseenter="stop" @mouseleave="up">
```

```
        <li v-for="value in news_list" :key="value">{{value}}</li>
      </ul>
    </div>
  </div>
</template>
<script>
import axios from 'axios'
export default {
  name: "MusicNews",
  data: function (){
    return {
      animate:false,                                    //是否使用指定类名
      timerID: null,                                    //定时器 ID
      news_list:[],                                     //最新资讯列表
    }
  },
  methods: {
    scrollUp: function (){
      var t = this;
      this.timerID = setInterval(function (){
        t.animate = true;                               //添加指定类名
        setTimeout(function (){
          t.news_list.push(t.news_list[0]);             //将数组第一个元素添加到数组末尾
          t.news_list.shift();                          //删除数组的第一个元素
          t.animate = false;                            //移除指定类名
        },500)
      }, 2000);
    },
    stop: function() {
      clearInterval(this.timerID);                      //停止向上滚动操作
    },
    up: function() {
      this.scrollUp();                                  //执行向上滚动操作
    },
    mounted : function(){
      axios.get('/data.json').then(function (response){ //发送 GET 请求
        this.news_list = response.data;                 //获取响应数据
      }.bind(this));
      this.scrollUp();                                  //执行向上滚动操作
    }
  }
}
</script>
```

> **说明**
> 为了能正常获取 data.json 文件中的数据,需要将该文件放在 public 文件夹中。

(2) 在 style.css 文件中编写 CSS 代码,为最新音乐资讯组件设置样式。代码如下:

```
.news{
    float: right;                    /*设置右浮动*/
    width:260px;                     /*设置宽度*/
    margin:0 auto;                   /*设置外边距*/
}
.news_title{
    font-size:18px;                  /*设置文字大小*/
}
.scroll{
    margin-left:5px;                 /*设置左外边距*/
    margin-top:5px;                  /*设置上外边距*/
    width:260px;                     /*设置宽度*/
    height:120px;                    /*设置高度*/
```

```
            overflow:hidden;                                  /*设置溢出内容隐藏*/
        }
        .scroll li{
            width:260px;                                      /*设置宽度*/
            height:30px;                                      /*设置高度*/
            line-height:30px;                                 /*设置行高*/
            margin-left:16px;                                 /*设置左外边距*/
        }
        .scroll li a{
            font-size:14px;                                   /*设置文字大小*/
            color:#333;                                       /*设置文字颜色*/
            text-decoration:none;                             /*设置链接文字无下画线*/
        }
        .scroll li a:hover{
            color:#66CCFF;                                    /*设置文字颜色*/
        }
        .anim{
            transition: all 0.5s;                             /*设置过渡效果*/
            margin-top: -30px;                                /*设置上外边距*/
        }
```

3.4.6 新歌首发

在歌曲列表展示界面的下方是新歌首发的展示界面。由于歌曲数量较多，因此采用分页的方式来展示首发歌曲的信息。具体的界面效果分别如图 3.9 和图 3.10 所示。

图 3.9　第一页新歌首发界面

图 3.10　第二页新歌首发界面

新歌首发界面的实现步骤如下。

（1）在 components 目录中创建 NewSong.vue 文件。在该文件的 <template> 标签中，调用 DataPage 组件，并将首发歌曲列表和每页显示的歌曲数量作为 Prop 属性进行传递给该组件。在 <script> 标签中，首先使用 import 引入 DataPage 组件，然后在 components 选项中注册该组件，并定义首发歌曲列表数组。代码如下：

```
<template>
    <DataPage :items = "dataList" :pageSize = "6" />
</template>
<script>
import DataPage from "@/components/DataPage.vue";
```

```js
export default {
  name: "NewSong",
  components: {
    DataPage
  },
  data: function (){
    return {
      dataList: [                                        //首发歌曲列表数组
        {
          imgUrl: require('@/assets/images/new/1.jpg'),  //歌曲图片 URL
          songName: 'Co-Pathetic',                       //歌曲名称
          singer: 'Novo Amor',                           //歌手名称
          duration: '03:55'                              //歌曲时长
        },
        {
          imgUrl: require('@/assets/images/new/2.jpg'),
          songName: 'Like That (Explicit)',
          singer: 'Future',
          duration: '04:27'
        },
        {
          imgUrl: require('@/assets/images/new/3.jpg'),
          songName: 'Run Run Run',
          singer: 'The Libertines',
          duration: '02:53'
        },
        {
          imgUrl: require('@/assets/images/new/4.jpg'),
          songName: 'Been Like This',
          singer: 'Meghan Trainor /T-Pain',
          duration: '02:25'
        },
        {
          imgUrl: require('@/assets/images/new/5.jpg'),
          songName: 'Team Side feat.RCB',
          singer: 'Alan Walker /Sofiloud',
          duration: '03:00'
        },
        {
          imgUrl: require('@/assets/images/new/6.jpg'),
          songName: 'Count Me Out',
          singer: 'Vicetone /Emily Falvey',
          duration: '03:13'
        },
        {
          imgUrl: require('@/assets/images/new/7.jpg'),
          songName: 'Still Here With You',
          singer: 'TheFatRat',
          duration: '02:28'
        },
        {
          imgUrl: require('@/assets/images/new/8.jpg'),
          songName: 'Higher Ground',
          singer: 'Purple Disco Machine',
          duration: '04:33'
        },
        {
          imgUrl: require('@/assets/images/new/9.jpg'),
          songName: 'You And Me',
          singer: 'Take That',
          duration: '03:34'
        },
        {
          imgUrl: require('@/assets/images/new/10.jpg'),
```

```
          songName: 'Lonely Cowboy',
          singer: 'KALEO',
          duration: '04:53'
        },
        {
          imgUrl: require('@/assets/images/new/11.jpg'),
          songName: 'Herobrine Phonk',
          singer: 'abtmelody',
          duration: '02:00'
        },
        {
          imgUrl: require('@/assets/images/new/12.jpg'),
          songName: 'The Last Bit Of Us',
          singer: 'Dean Lewis',
          duration: '03:22'
        }
      ]
    }
  }
}
</script>
```

（2）在 components 目录中创建 DataPage.vue 文件。在该文件的<template>标签中，使用 v-for 指令遍历当前页的首发歌曲列表信息，包括歌曲图片、歌曲名称、歌手名称和歌曲时长。同时，添加"<"和">"按钮进行分页控制。单击"<"按钮会调用 prevPage()方法，单击">"按钮会调用 nextPage()方法。在<script>标签中定义父组件传递的 Prop 属性、数据、计算属性和方法。totalPages 计算属性用于获取总页数，currentPageItems 计算属性用于获取当前页的首发歌曲列表，prevPage()方法和 nextPage()方法分别实现当前页减 1 和加 1 的操作。代码如下：

```
<template>
  <div class="page">
    <div class="page_title">新歌首发</div>
    <div class="page_item" v-for="item in currentPageItems" :key="item">
      <div class="img"><img :src="item.imgUrl"></div>
      <div class="name">
        <div>{{item.songName}}</div>
        <div>{{item.singer}}</div>
      </div>
      <div class="duration">{{item.duration}}</div>
    </div>
    <div class="page_control">
      <button @click="prevPage" :disabled = "currentPage === 1">&lt;</button>
      <span>{{currentPage}}/{{totalPages}}</span>
      <button @click="nextPage" :disabled = "currentPage === totalPages">&gt;</button>
    </div>
  </div>
</template>
<script>
export default {
  name: "DataPage",
  props: ['items','pageSize'],
  data: function (){
    return {
      currentPage: 1                                         //当前页
    }
  },
  computed: {
    totalPages(){                                            //总页数
      return Math.ceil(this.items.length / this.pageSize);
    },
    currentPageItems(){                                      //当前页数据
```

```
        let start = (this.currentPage - 1) * this.pageSize;
        let end = start + this.pageSize;
        return this.items.slice(start, end);
      }
    },
    methods: {
      prevPage(){
        if(this.currentPage > 1) this.currentPage -= 1;         //当前页减1
      },
      nextPage(){
        if(this.currentPage < this.totalPages) this.currentPage += 1;   //当前页加1
      }
    }
  }
</script>
```

（3）在 style.css 文件中编写 CSS 代码，为新歌首发组件设置样式。代码如下：

```
.page{
    clear: both;                                    /*清除浮动*/
}
.page .page_title{
    height: 80px;                                   /*设置高度*/
    line-height: 80px;                              /*设置行高*/
    font-size:28px;                                 /*设置文字大小*/
    letter-spacing: 10px;                           /*设置文字间距*/
    text-align: center;                             /*设置文本水平居中显示*/
}
.page .page_item{
    float: left;                                    /*设置左浮动*/
    width: 33%;                                     /*设置宽度*/
    height: 100px;                                  /*设置高度*/
}
.page img{
    width: 80px;                                    /*设置宽度*/
    height: 80px;                                   /*设置高度*/
}
.page .page_item>div{
    float: left;                                    /*设置左浮动*/
}
.page .page_item .name{
    margin-top: 12px;                               /*设置上外边距*/
    margin-left: 15px;                              /*设置左外边距*/
}
.page .page_item .name div{
    margin-top: 5px;                                /*设置上外边距*/
}
.page .page_item .duration{
    float: right;                                   /*设置右浮动*/
    margin-top: 30px;                               /*设置上外边距*/
    margin-right: 20px;                             /*设置右外边距*/
}
.page .page_control{
    clear: both;                                    /*清除浮动*/
    text-align: center;                             /*设置文本水平居中显示*/
    margin-bottom: 10px;                            /*设置下外边距*/
}
.page .page_control span{
    margin: 0 10px;                                 /*设置外边距*/
}
.page .page_control button{
    width: 60px;                                    /*设置宽度*/
    height: 30px;                                   /*设置高度*/
}
```

3.4.7 首页底部的设计

网站首页的底部展示了版权等方面的信息。首页底部的效果如图 3.11 所示。

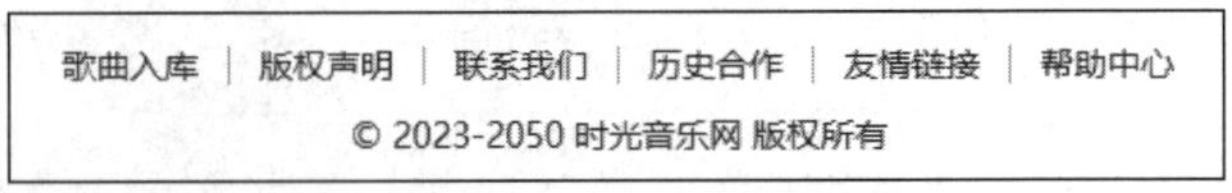

图 3.11 首页底部效果

首页底部的实现步骤如下。

（1）在 components 目录中创建 MusicBottom.vue 文件。在该文件的<template>标签中，定义 ul 列表，并在该列表中使用 v-for 指令对版权信息列表 b_list 进行遍历。在<script>标签中，定义版权信息列表 b_list。具体代码如下：

```
<template>
  <div class="bottom">
    <ul>
      <li v-for="value in b_list" :key="value">{{value}}</li>
    </ul>
    <div class="copyright">© 2023-2050 时光音乐网 版权所有</div>
  </div>
</template>
<script>
export default {
  name: "MusicBottom",
  data: function (){
    return {
      b_list: ['歌曲入库','版权声明','联系我们','历史合作','友情链接','帮助中心']
    }
  }
}
</script>
```

（2）在 style.css 文件中编写 CSS 代码，为页面底部组件设置样式。代码如下：

```
.bottom{
    width: 1000px;                           /*设置宽度*/
    height: 80px;                            /*设置高度*/
    margin: 0 auto;                          /*设置外边距*/
    border-top: 1px solid #3366FF;           /*设置上边框*/
    text-align: center;                      /*设置文本水平居中显示*/
}
.bottom ul{
    padding-top: 15px;                       /*设置上内边距*/
    margin-bottom: 10px;                     /*设置下外边距*/
}
.bottom li{
    display: inline-block;                   /*设置行内块元素*/
    width: 80px;                             /*设置宽度*/
    height: 16px;                            /*设置高度*/
    line-height: 16px;                       /*设置行高*/
    text-align: center;                      /*设置文本水平居中显示*/
}
.bottom li:not(:last-child){
    border-right: 1px solid #BBBBBB;         /*设置右边框*/
}
```

3.4.8 在根组件中构建音乐网首页

修改 App.vue 文件，首先在<script>标签中使用 import 引入构建音乐网首页的多个组件，然后在 components 选项中注册各个组件，并在<template>标签中调用各个组件，最后在<style>标签中引入公共 CSS

文件 style.css。代码如下：

```html
<template>
  <div id="app">
    <div class="logo"><img src="./assets/images/ball.png"></div>
    <MusicNav/>
    <div class="main">
      <SongList/>
      <MusicBanner/>
      <MusicTabs/>
      <MusicNews/>
      <NewSong/>
    </div>
    <MusicBottom/>
  </div>
</template>
<script>
import MusicNav from './components/MusicNav.vue'
import SongList from './components/SongList.vue'
import MusicBanner from './components/MusicBanner.vue'
import MusicTabs from './components/MusicTabs.vue'
import MusicNews from './components/MusicNews.vue'
import NewSong from "@/components/NewSong.vue";
import MusicBottom from './components/MusicBottom.vue'
export default {
  name: 'App',
  components: {
    MusicNav,          //网站导航栏组件
    SongList,          //歌曲列表组件
    MusicBanner,       //轮播图组件
    MusicTabs,         //选项卡组件
    MusicNews,         //最新资讯组件
    NewSong,           //新歌首发组件
    MusicBottom        //页面底部组件
  }
}
</script>
<style lang="scss">
@import 'assets/css/style.css';
</style>
```

3.5 项目运行

通过前述步骤，我们已经设计并完成了"时光音乐网首页设计"项目的开发。接下来，我们运行该项目，以检验我们的开发成果。首先打开命令提示符窗口，切换到项目所在目录，然后执行 npm run serve 命令运行该项目，如图 3.12 所示。

在浏览器地址栏中输入 http://localhost:8080 并按 Enter 键后，即可成功运行该项目，具体效果如图 3.13 所示。

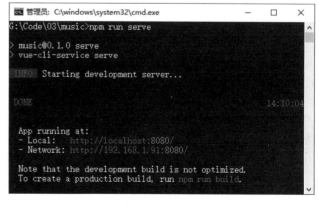

图 3.12　运行项目

图 3.13 时光音乐网首页设计效果

3.6 源码下载

本章虽然详细地讲解了如何编码实现"时光音乐网首页设计"的各个功能，但给出的代码都是代码片段，而非完整的源代码。为了方便读者学习，本书提供了该项目的完整源代码，读者可以扫描右侧的二维码进行下载。

第 4 章 游戏公园博客

——Vue CLI + Vue Router + Vuex

互联网时代,电子游戏已经成为人们生活中不可或缺的一部分。当今电子游戏种类繁多,根据媒介的不同,可以分为五大类:电脑游戏、主机游戏(或称家用机游戏、电视游戏)、掌机游戏、街机游戏和移动游戏(主要是手机游戏)。本章将以电子游戏为主题,使用 Vue CLI、Vue Router、Vuex 等技术设计一个电子游戏资讯网站——游戏公园博客。

项目微视频

本项目的核心功能及实现技术如下:

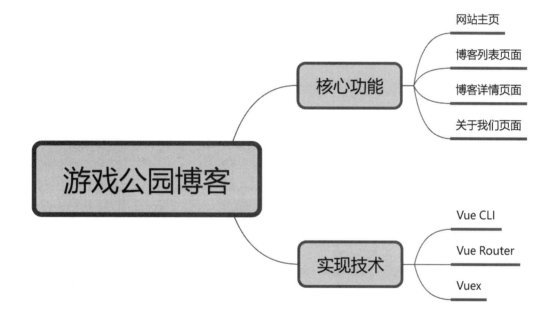

4.1 开发背景

随着互联网的快速发展,网络游戏已经成为一个热门的话题。由于网络游戏的形式丰富多样,内容广泛,网络游戏已经成为人们减缓压力、调节情绪的一种重要选择。随着计算机的普及,人们更愿意使用互联网来获取游戏方面的信息。互联网能够让人们简便快捷、足不出户地满足对游戏信息的需求。本章将使用 Vue CLI、Vue Router、Vuex 等开发一个游戏资讯类网站,其实现目标如下:

- ☑ 在主页中展示推荐游戏和最新游戏。
- ☑ 实现游戏博客列表页面。
- ☑ 在博客详情页面显示游戏介绍。
- ☑ 用户可以在博客详情页面发表游戏评论。
- ☑ 在关于我们页面展示网站制作团队信息。

4.2 系统设计

4.2.1 开发环境

本项目的开发及运行环境如下：
- ☑ 操作系统：推荐 Windows 10、Windows 11 或更高版本，同时兼容 Windows 7（SP1）。
- ☑ 开发工具：WebStorm。
- ☑ 开发框架：Vue.js 3.0。

4.2.2 业务流程

游戏公园网站由多个页面组成，包括主页、博客列表页面、博客详情页面和关于我们页面。根据该网站的业务需求，我们设计如图 4.1 所示的业务流程图。

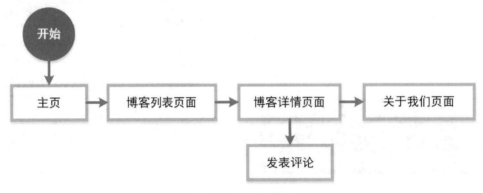

图 4.1　业务流程图

4.2.3 功能结构

本项目的功能结构已经在章首页中给出。其实现的具体功能如下：
- ☑ 主页：包含"推荐游戏"和"最新游戏"两个子功能，为用户推荐和介绍最新的游戏资讯。
- ☑ 博客功能：分成"博客列表"和"博客详情"两个子功能，方便用户查找和浏览游戏资讯。
- ☑ 发表评论：用户可以发表游戏评论，并在页面中展示评论内容。
- ☑ 关于我们：介绍游戏公园网站的发展历史和网站特点等。

4.3 技术准备

在开发游戏公园网站时，我们主要应用 Vue CLI 脚手架工具、Vue Router、Vuex。其中，Vue CLI 已经在第 3 章中做了介绍，下面将针对本项目所用的 Vue Router 和 Vuex 技术点及其具体作用进行简述。

1. Vue Router

Vue Router 是 Vue.js 官方的路由管理器。在单页 Web 应用中，整个项目只有一个 HTML 文件，不同视图（组件的模板）的内容都是在同一个页面中动态渲染的。路由的作用是当用户切换页面时，实现页面之间的跳转。

在使用 Vue Router 之前需要对其进行安装。如果在项目中使用 Vue Router，则可以使用 npm 方式进行安装。在命令提示符窗口中输入命令如下：

```
npm install vue-router@next --save
```

Vue.js 路由的思想是通过不同的 URL 访问不同的内容。要想通过路由实现组件之间的切换，需要使用 Vue Router 提供的<router-link>组件。该组件用于创建一个导航链接，通过设置 to 属性链接到一个目标地址，从而切换显示不同的内容。切换后的内容将在页面上通过<router-link>组件指定的位置进行动态渲染。

2. Vuex

Vuex 是一个专门为 Vue.js 应用程序开发的状态管理模式。它采用集中式存储来管理应用程序中所有组件的状态。在通常情况下，每个组件都拥有自己的状态。有时，需要将某个组件的状态变化传递给其他组件，以使它们也进行相应的修改。这时可以使用 Vuex 保存需要管理的状态值，一旦这些值被修改，所有引用这些值的组件就会自动更新。

在使用 Vuex 之前需要对其进行安装。如果使用模块化开发，可以通过 npm 进行安装。在命令提示符窗口中输入以下命令：

```
npm install vuex@next --save
```

Vuex 主要由 5 部分组成，分别为 state、getters、mutations、actions 和 modules。在 state 中可以定义需要共享的数据；getters 用于对状态进行一些处理，类似于 Vue 实例中的 computed 选项；mutations 用于存储更改 state 状态的方法，是 Vuex 中唯一修改 state 的方式，但不支持异步操作，类似于 Vue 实例中的 methods 选项；actions 可以通过提交 mutations 中的方法来改变状态，支持异步操作；modules 是 store 的子模块，内容相当于 store 的一个实例。

> **说明**
> 如果使用 Vue CLI 创建项目，则可以选择手动对项目进行配置，在项目的配置选项中应用空格键选择 Router 和 Vuex。这样，在创建项目后，会自动安装 Vue Router 和 Vuex，无须进行单独安装。

有关 Vue CLI、Vue Router、Vuex 等基础知识，《Vue.js 从入门到精通》一书中进行了详细的讲解。对这些知识尚不熟悉的读者，建议参考该书的相关章节进行深入学习。

4.4 创建项目

在设计游戏公园博客各功能模块之前，需要使用 Vue CLI 创建项目，并将项目名称设置为 gamepark。在命令提示符窗口中输入以下命令：

```
vue create gamepark
```

按 Enter 键，选择 Manually select features，如图 4.2 所示。

接着按 Enter 键，选择 Router 和 Vuex 选项，如图 4.3 所示。

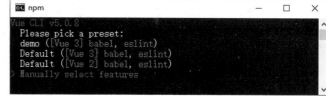

图 4.2　选择 Manually select features

然后选择路由是否使用 history 模式，输入 y 表示使用 history 模式，如图 4.4 所示。

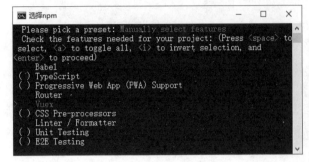

图 4.3　选择配置选项

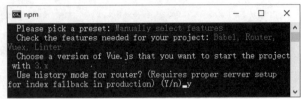

图 4.4　使用 history 模式

创建项目后，先整理项目的目录，然后在 assets 目录中分别创建 css、images 和 fonts 三个文件夹。其中：在 css 文件夹中创建 style.css 文件，该文件将被作为网站的公共样式文件；在 images 文件夹中存储网站所需的所有图片。在 fonts 文件夹中存储字体文件。完成这些准备工作之后，下面开始设计网站的各个组成页面。

4.5　功 能 设 计

4.5.1　主页设计

主页主要包含 4 个部分，分别是头部区域、推荐游戏区域、最新游戏区域和底部区域。在该项目中：各页面的头部区域和底部区域是相同的；推荐游戏区域通过轮播图的形式显示游戏内容。主页的头部区域和推荐游戏区域的效果如图 4.5 所示。

图 4.5　主页的头部区域和推荐游戏区域

最新游戏区域通过分页的形式显示游戏列表，当鼠标悬停在游戏图片上时，会以动画的形式显示游戏简介。主页的最新游戏区域和底部区域的效果如图 4.6 所示。

图 4.6　主页的最新游戏区域和底部区域的效果

主页的实现过程如下。

（1）在 components 文件夹下创建页面头部组件 TheHead.vue 和页面底部组件 TheFooter.vue。在 TheHead.vue 文件中，当单击不同的超链接时会跳转到相应的页面。代码如下：

```
<template>
  <div class="header" >
    <div class="header-top">
      <div class="container">
        <div class="head-top">
          <div class="logo">
            <h1><a @click="show('index')">游戏<span>公园</span></a></h1>
          </div>
          <div class="top-nav">
            <span class="menu"><img src="../assets/images/menu.png"> </span>
            <ul>
              <li class="active"><a @click="show('index')">主页</a></li>
              <li><a @click="show('blog')">博客</a></li>
              <li><a @click="show('about')">关于我们</a></li>
              <div class="clearfix"> </div>
            </ul>
          </div>
          <div class="clearfix"> </div>
        </div>
      </div>
    </div>
  </div>
</template>
<script>
export default {
  name: "TheHead",
  methods: {
    show: function (value){
      this.$router.push({name: value});         //页面跳转
    }
```

```
    }
  }
</script>
```

TheFooter.vue 组件的代码如下：

```
<template>
  <div class="footer">
    <div class="container">
      <ul class="footer-grid">
        <li class="active"><a @click="show('index')">主页</a></li>
        <li><a @click="show('blog')">博客</a></li>
        <li><a @click="show('about')">关于我们</a></li>
      </ul>
      <p>
        游戏公园　|　设计 by　<a href="http://www.mingrisoft.com/" target="_blank">吉林省明日科技有限公司</a>
      </p>
    </div>
  </div>
</template>
<script>
export default {
  name: "TheFooter",
  methods: {
    show: function (value){
      this.$router.push({name: value});
    }
  }
}
</script>
```

（2）在 views 文件夹下创建主页文件夹 index，并在 index 文件夹下创建推荐游戏区域组件 TheBanner.vue。该组件主要用于实现推荐游戏图片轮播的动画效果。游戏图片下方有 3 个数字按钮，用户单击不同的按钮将调用 scrollLeft()方法，从而实现图片向左或向右滚动的效果。当鼠标移入图片时，stop()方法会被调用，该方法可以使图片停止滚动；当鼠标移出图片时，goLeft()方法会被调用，该方法可以使图片继续滚动。TheBanner.vue 组件的代码如下：

```
<template>
  <div class="banner">
    <div class="container">
      <h2>推荐游戏</h2>
      <div class="banner-matter">
        <div class="slider">
          <div class="bd">
            <div id="wrap" :style="{left:dis}" @mouseenter="stop" @mouseleave="goLeft">
              <ul id="slider">
                <li v-for="imgUrl in imgUrlArr" :key="imgUrl">
                  <img width="246" :src="imgUrl">
                </li>
              </ul>
            </div>
          </div>
          <div id="opts">
            <span :class="{active: n===0}" @click="scrollLeft(0)">1</span>
            <span :class="{active: n===1}" @click="scrollLeft(1)">2</span>
            <span :class="{active: n===2}" @click="scrollLeft(2)">3</span>
          </div>
        </div>
      </div>
    </div>
  </div>
</template>
<script>
```

```
export default {
  name: "TheBanner",
  data: function (){
    return {
      dis: 0,                                                  //向左滚动距离
      timerID: null,                                           //定时器 ID
      n: 0,                                                    //数字按钮索引
      flag: true,                                              //控制单击数字按钮
      active: true,                                            //激活按钮样式
      imgUrlArr: [                                             //图片 URL 数组
        require('@/assets/images/recommend/ta1.jpg'),
        require('@/assets/images/recommend/ta2.jpg'),
        require('@/assets/images/recommend/ta3.jpg'),
        require('@/assets/images/recommend/ta4.jpg'),
        require('@/assets/images/recommend/ta5.jpg'),
        require('@/assets/images/recommend/ta6.jpg'),
        require('@/assets/images/recommend/ta7.jpg'),
        require('@/assets/images/recommend/ta8.jpg'),
        require('@/assets/images/recommend/ta9.jpg'),
        require('@/assets/images/recommend/ta10.jpg'),
        require('@/assets/images/recommend/ta11.jpg'),
        require('@/assets/images/recommend/ta12.jpg')
      ]
    }
  },
  methods: {
    scrollLeft: function(n){
      if(this.flag){                                           //当 flag 为 true 时，单击数字按钮实现图片滚动
        this.flag = false;                                     //当 flag 为 false 时，单击数字按钮无效
        this.n = n;
        clearInterval(this.timerID);                           //取消定时器
        this.autoScroll();                                     //图片自动滚动
        this.dis = -this.n * 1000 + 'px';                      //设置图片滚动距离
        var t = this;
        setTimeout(function (){
          t.flag = true;                                       //过一秒后，可以继续单击数字按钮
        },1000);
      }
    },
    autoScroll: function (){
      var t = this;
      this.timerID = setInterval(function (){
        t.n === 2 ? t.n = 0 : t.n++;
        t.dis = -t.n * 1000 + 'px';                            //设置图片滚动距离
      },2000);
    },
    stop: function (){
      clearInterval(this.timerID);                             //取消定时器，图片停止滚动
    },
    goLeft: function (){
      this.autoScroll();                                       //图片自动滚动
    }
  },
  mounted: function (){
    this.autoScroll();                                         //图片自动滚动
  }
}
</script>

<style scoped>
  ul {
    list-style:none;                                           /*设置列表无样式*/
  }
  .bd{
```

```css
    width:1000px;                                    /*设置宽度*/
    height: 360px;                                   /*设置高度*/
    position: relative;                              /*设置相对定位*/
    overflow: hidden;                                /*设置溢出内容隐藏*/
    margin:30px auto;                                /*设置外边距*/
}
#wrap {
    position: absolute;                              /*设置绝对定位*/
    top: 0;                                          /*设置元素到父元素顶部的距离*/
    left: 0;                                         /*设置元素到父元素左端的距离*/
    transition: all 1s;                              /*设置过渡效果*/
}
#slider {
    width:400%;                                      /*设置宽度*/
    font:100px/400px Microsoft Yahei;                /*设置字体*/
    text-align:center;                               /*设置文本水平居中显示*/
    color:#fff;                                      /*设置文字颜色*/
}
#slider li {
    float:left;                                      /*设置左浮动*/
    width:250px;                                     /*设置宽度*/
}
#opts {
    width:600px;                                     /*设置宽度*/
    height:40px;                                     /*设置高度*/
    margin-left:500px;                               /*设置左外边距*/
    color:#fff;                                      /*设置文字颜色*/
    text-align:center;                               /*设置文本水平居中显示*/
    font:20px/40px Microsoft Yahei;                  /*设置字体*/
}
#opts span {
    float:left;                                      /*设置左浮动*/
    width:50px;                                      /*设置宽度*/
    height:40px;                                     /*设置高度*/
    margin-right:4px;                                /*设置右外边距*/
    background:#01254a;                              /*设置背景颜色*/
    cursor:pointer;                                  /*设置鼠标形状*/
    border-radius: 50px;                             /*设置圆角边框*/
}
#opts span:hover {
    background:#405871;                              /*设置背景颜色*/
}
#opts span.active{
    background:#3399FF;                              /*设置背景颜色*/
}
</style>
```

（3）在 index 文件夹下创建最新游戏区域组件 TheContainer.vue。在 TheContainer.vue 组件的<template>标签中，调用 DataPage 组件，并将游戏图片的 URL 数组以及每页显示的图片数量作为 Prop 属性传递给 DataPage 组件。在<script>标签中使用 import 引入 DataPage 组件，然后在 components 选项中注册该组件，接着定义图片的 URL 数组。代码如下：

```
<template>
    <DataPage :items="imgUrlArr" :pageSize="6"/>
</template>
<script>
import DataPage from "@/views/index/DataPage.vue";
export default {
  name: "TheContainer",
  components: {
    DataPage
  },
  data: function (){
```

```
    return {
      imgUrlArr: [                                              //图片 URL 数组
        require('@/assets/images/new/a1.jpg'),
        require('@/assets/images/new/a2.jpg'),
        require('@/assets/images/new/a3.jpg'),
        require('@/assets/images/new/a4.jpg'),
        require('@/assets/images/new/a5.jpg'),
        require('@/assets/images/new/a6.jpg'),
        require('@/assets/images/new/a7.jpg'),
        require('@/assets/images/new/a8.jpg'),
        require('@/assets/images/new/a9.jpg'),
        require('@/assets/images/new/a10.jpg'),
        require('@/assets/images/new/a11.jpg'),
        require('@/assets/images/new/a12.jpg')
      ]
    }
  }
}
</script>
```

（4）在 index 文件夹下创建分页组件 DataPage.vue。该组件主要用来分页展示最新游戏列表，当鼠标悬停在游戏图片上时，以动画的形式显示游戏简介，代码如下：

```
<template>
  <div class="container">
    <div class="games">
      <h3>最新游戏</h3>
      <section>
        <ul id="da-thumbs" class="da-thumbs">
          <li v-for="url in currentPageData" :key="url">
            <a @click="show('single')" rel="title" class="b-link-stripe b-animate-go thickbox"
               @mouseenter="mouseEnter" @mouseleave="mouseLeave" >
              <img :src="url">
              <div style="left: -100%; display: block; top: 0; transition: all 300ms ease;">
                <h5>Games</h5>
                <span>领先的在线休闲游戏平台</span>
              </div>
            </a>
          </li>
          <div class="clearfix"> </div>
        </ul>
        <div class="page_control">
          <button @click="prevPage" :disabled = "currentPage === 1"><img src="@/assets/images/left.png"></button>
          <span>{{currentPage}}/{{totalPages}}</span>
          <button @click="nextPage" :disabled = "currentPage === totalPages">
            <img src="@/assets/images/right.png">
          </button>
        </div>
      </section>
    </div>
  </div>
</template>
<script>
export default {
  name: "DataPage",
  props: [
    'items',
    'pageSize'
  ],
  data: function (){
    return {
      currentPage: 1                                            //当前页
    }
  },
```

```
    computed: {
      totalPages(){                                                    //总页数
        return Math.ceil(this.items.length / this.pageSize);
      },
      currentPageData(){                                               //当前页数据
        let start = (this.currentPage - 1) * this.pageSize;
        let end = start + this.pageSize;
        return this.items.slice(start, end);
      },
    },
    methods: {
      mouseEnter: function (event){                                    //鼠标滑入
        var menu=event.target.lastElementChild;                        //获取对象最后的子元素节点
        menu.style.left='0px';                                         //将节点的样式属性 left 值设置为 0px
      },
      mouseLeave: function (event){                                    //鼠标滑出
        var menu=event.target.lastElementChild;                        //获取对象最后的子元素节点
        menu.style.left='-100%';                                       //将节点的样式属性 left 值设置为-100%
      },
      show: function (value){
        this.$router.push({name: value});
      },
      prevPage(){
        if(this.currentPage > 1) this.currentPage -= 1;                //当前页减 1
      },
      nextPage(){
        if(this.currentPage < this.totalPages) this.currentPage += 1;  //当前页加 1
      }
    }
  }
</script>
<style scoped>
.games .page_control{
  text-align: center;                                                  /*设置文本水平居中显示*/
  margin-top: 10px;                                                    /*设置下外边距*/
}
.games .page_control span{
  margin: 0 10px;                                                      /*设置外边距*/
  font-size: 16px;
}
.games .page_control button{
  background-color: #333;
  width: 60px;                                                         /*设置宽度*/
  height: 50px;                                                        /*设置高度*/
  border: none;                                                        /*设置无边框*/
}
.games .page_control button:disabled{
  background-color: #ccc;                                              /*设置背景颜色*/
}
</style>
```

（5）在 index 文件夹下创建 IndexHome.vue 组件。该组件是 TheBanner 和 TheContainer 组件的父组件，代码如下：

```
<template>
  <div>
    <TheBanner/>
    <TheContainer/>
  </div>
</template>
<script>
import TheBanner from "./TheBanner";
import TheContainer from "./TheContainer";
```

```
export default {
  name: "IndexHome",
  components: {
    TheBanner,
    TheContainer
  }
}
</script>
```

4.5.2 博客列表页面设计

博客列表是游戏公园博客资讯平台的核心功能。该列表主要展示相关游戏的名称、缩略图、游戏简介，以及"更多信息"按钮。单击"更多信息"按钮，页面会跳转到博客详情页面。博客列表页面的效果如图 4.7 所示。

图 4.7　博客列表页面效果

博客列表页面的具体实现方法如下。

在 views 文件夹下创建 blog 文件夹，在该文件夹下创建博客列表组件 BlogHome.vue。在该组件中，我们将游戏名称、缩略图和游戏简介等信息定义在数组中，并使用 v-for 指令遍历该数组，代码如下：

```html
<template>
  <div class="blog">
    <div class="container">
      <h3>博客</h3>
      <div class="blog-head">
        <div class="col-md-4 blog-top" v-for="blog in blogInfo" :key="blog">
          <div class="blog-in">
            <a @click="show('single')" target="_blank"><img class="img-responsive" :src="blog.url"></a>
            <div class="blog-grid">
              <h4><a @click="show('single')">{{ blog.name }}</a></h4>
              <p>{{ blog.intro }}</p>
              <div class="date">
                <span class="date-in"><i class="glyphicon glyphicon-calendar"></i>{{ blog.date }}</span>
                <a @click="show('single')" class="comments">
                  <i class="glyphicon glyphicon-comment"></i>{{ blog.count }}
                </a>
                <div class="clearfix"> </div>
              </div>
              <div class="more-top">
                <a class=" hvr-wobble-top" @click="show('single')">更多信息</a>
              </div>
            </div>
          </div>
        </div>
        <div class="clearfix"> </div>
      </div>
    </div>
  </div>
</template>
<script>
export default {
  name: "AboutHome",
  data: function (){
    return {
      blogInfo: [
        {
          url: require('@/assets/images/b1.jpg'),
          name: '超凡蜘蛛侠 2',
          intro: `《超凡蜘蛛侠 2 The Amazing Spider-Man 2》是一款 Gameloft 出品的动作游戏,
            又名《蜘蛛人驚奇再起 2》。游戏中,你将化身为惊奇蜘蛛人,
            面对这位蛛丝射手最大的挑战！......`,
          date: '22.10.2024',
          count: 90
        },
        {
          url: require('@/assets/images/b2.jpg'),
          name: '神庙逃亡',
          intro: `《神庙逃亡》是由 Imangi Studios 开发制作的一款跑酷冒险类单机类系列游戏。
            游戏内容和大多数跑酷游戏都非常相似,越过重重障碍和陷阱,不断向前飞奔。让我们一起出发吧！`,
          date: '25.10.2024',
          count: 76
        },
        {
          url: require('@/assets/images/b3.jpg'),
          name: '地铁跑酷(周年庆)',
          intro: `全球超人气跑酷手游《地铁跑酷》给你精彩、好玩的游戏体验。画面精致、操作流畅、
            玩法刺激、滑板炫酷、角色丰富、特效绚丽……全民皆玩,全球 3 亿用户的共同选择,
            一路狂奔,环游世界,你会爱上它！`,
          date: '22.11.2024',
          count: 36
        },
        {
          url: require('@/assets/images/b4.jpg'),
```

```
          name: '蝙蝠侠',
          intro: `《蝙蝠侠 Batman》是一款以蝙蝠侠为主题的动作游戏，游戏里玩家将化身为蝙蝠侠维护哥谭市的治安。
                不同于其他英雄人物，玩家除了惩治坏人，还需要为哥谭市的未来做出选择……`,
          date: '26.10.2024',
          count: 63
        },
        {
          url: require('@/assets/images/b5.jpg'),
          name: '愤怒的小鸟',
          intro: `《愤怒的小鸟(6周年版)》是一款风靡全球的物理解谜游戏。游戏中小鸟们为了报复偷走鸟蛋的猪，
                鸟儿以自己的身体为武器，攻击猪们的堡垒。花好月圆夜鸟出没。开始游戏吧！`,
          date: '02.10.2024',
          count: 57
        },
        {
          url: require('@/assets/images/b6.jpg'),
          name: '时空召唤(新英雄马超)',
          intro: `《时空召唤》是一款以科幻为题材的竞技 MOBA 手游，由银汉游戏倾力打造，是《时空猎人》兄弟产品。
                不缩水的 5V5 大地图，坚持平衡的游戏规则，原创的科幻英雄，纯粹的 MOBA 玩法。`,
          date: '27.10.2024',
          count: 56
        }
      ]
    }
  },
  methods: {
    show: function (value){
      this.$router.push({name: value});
    }
  }
}
</script>
```

4.5.3　博客详情页面设计

用户在博客列表页面单击游戏图片、游戏名称或"更多信息"按钮时，就可以进入博客详情页面。博客详情页面主要包括游戏介绍和发表评论两部分内容。游戏介绍部分的界面效果如图 4.8 所示。

图 4.8　游戏介绍部分的界面效果

在游戏介绍界面的下方提供了发表游戏评论的功能。用户输入评论信息，单击"提交"按钮就可以发表评论。图 4.9 和图 4.10 分别展示了游戏评论输入界面和游戏评论提交后的效果。

图 4.9　游戏评论输入界面效果　　　　　图 4.10　游戏评论提交后的界面效果

具体实现步骤如下。

（1）在 views 文件夹下创建 single 文件夹，在该文件夹下创建博客详情组件 SingleHome.vue。在 SingleHome.vue 组件的<template>标签中添加游戏介绍和用户发表评论的表单。在<script>标签中引入 mapState()和 mapMutations()辅助函数，以便实现组件中的计算属性、方法和 store 中的 state、mutation 之间的映射。用户输入评论内容后，单击"提交"按钮会调用 save()方法，在该方法中将用户发表的评论信息作为参数传递给 store 实例的 mutation，以实现保存评论信息的功能。具体代码如下：

```html
<template>
  <div class="container">
    <div class="single">
      <a href="#"><img class="img-responsive" src="../../assets/images/si.jpg"></a>
      <div class="single-grid" style="font-size: 16px">
        <h4>地铁跑酷(周年庆) </h4>
        <div class="cal">
          <ul>
            <li><span><i class="glyphicon glyphicon-calendar"> </i>2024/11-22</span></li>
            <li><a href="#"><i class="glyphicon glyphicon-comment"></i>36</a></li>
          </ul>
        </div>
        <p>全球超人气跑酷手游《地铁跑酷》给你精彩、好玩的游戏体验。画面精致、操作流畅、
          玩法刺激、滑板炫酷、角色丰富、特效绚丽……全民皆玩，全球 3 亿用户的共同选择，
          一路狂奔，环游世界，你会爱上它！
        </p>
        <p>更新提示<br/>
          1.圣诞节快乐！尽情享受圣诞节在雪中参加地铁跑酷的乐趣；<br/>
          2.欢迎极地探索者——马利克和他的长牙装扮；<br/>
          3.沿着奇妙的玩具工厂滑板冲浪，探索美丽的冰雪洞窟；<br/>
          4.来和拥有着冰雪装扮的精灵琪琪一起玩耍吧；<br/>
          5.在新的冰川滑板上滑雪冲浪！<br/>
        </p>
      </div>
      <div class="comments-top">
        <h3>评论</h3>
        <div class="media" v-for="item in comment" :key="item">
          <div class="media-left">
            <a href="#">
              <img :src="item.img"> </a>
```

```html
        </div>
        <div class="media-body">
          <h4 class="media-heading">{{ item.name }}</h4>
          <p>{{ item.content }}</p>
        </div>
      </div>
    </div>
    <div class="comment-bottom">
      <h3>发表评论</h3>
      <form>
        <input type="text" placeholder="姓名" v-model="name">
        <textarea type="text" placeholder="内容" required v-model="content"></textarea>
        <input type="submit" value="提交" @click="save">
      </form>
    </div>
  </div>
</div>
</template>
<script>
```

```javascript
import {mapState,mapMutations} from 'vuex'          //引入 mapState 和 mapMutations
export default {
  name: "SingleHome",
  data: function (){
    return {
      img : require("@/assets/images/si.png"),
      name: '',
      content: ''
    }
  },
  computed: {
    ...mapState(['comment'])
  },
  methods: {
    ...mapMutations(['add']),
    save: function (){
      this.name = this.name === '' ? '匿名用户' : this.name;    //评论用户名
      if(this.content === '') return;
      //调用 add()方法添加用户评论
      this.add({img: this.img,name: this.name,content: this.content});
    }
  }
}
</script>
```

（2）修改 store 文件夹下的 index.js 文件，在 store 实例中分别定义 state 和 mutation。当用户发表评论后，使用 localStorage.setItem 存储添加评论信息后的评论列表。具体代码如下：

```javascript
import { createStore } from 'vuex'

export default createStore({
  state: {
    comment: localStorage.getItem('list')?JSON.parse(localStorage.getItem('list')):[{
      img : require("@/assets/images/si.png"),
      name : "Tony",
      content: '游戏玩法操作简单，是休闲娱乐的不错选择。'
    },{
      img : require("@/assets/images/si.png"),
      name : "Kelly",
      content: '玩家可以通过积攒金币和钥匙来解锁角色，体验很棒。'
    }]
  },
  mutations: {
```

```
    add: function (state,newComment){                              //添加用户评论
      state.comment.push(newComment);
      localStorage.setItem('list', JSON.stringify(state.comment));
    }
  }
})
```

4.5.4 关于我们页面设计

关于我们页面主要用来展示网站的特色版块、特色功能和制作团队信息。关于我们页面的效果如图 4.11 所示。

图 4.11 关于我们页面的效果

关于我们页面的具体实现方法如下。

在 views 文件夹下创建 about 文件夹，在该文件夹下创建关于我们页面组件 AboutHome.vue。在 AboutHome.vue 组件的 <template> 标签中，首先添加关于我们页面的文字简述、图片以及特色板块和特色功能，然后实现团队信息展示的功能，并使用 v-for 指令遍历团队人员信息，包括团队人员的头像、名称和职业。代码如下：

```
<template>
  <div class="about">
    <div class="container">
      <div class="about-top">
        <h3>关于我们</h3>
      </div>
      <div class="about-bottom">
        <p style="text-align: left;font-size: 20px">
          <strong>
            明日学院，是吉林省明日科技有限公司倾力打造的在线实用技能学习平台，该平台于2016年正式上线，
            主要为学习者提供海量、优质的课程，课程结构严谨，用户可以根据自身的学习程度，自主安排学习进度。
            我们的宗旨是，为编程学习者提供一站式服务，培养用户的编程思维。
          </strong>
        </p>
        <div class="about-btm">
          <div class="col-md-6 about-left">
            <a @click="show('single')"><img class="img-responsive" src="../../assets/images/bt.jpg" alt=""/></a>
          </div>
          <div class="col-md-6 about-right">
            <a @click="show('single')"><img class="img-responsive" src="../../assets/images/bt1.jpg" alt=""/></a>
          </div>
          <div class="clearfix"></div>
        </div>
      </div>
      <!--advantages-->
      <div class="advantages">
        <div class="col-md-6 advantages-left ">
          <h3>特色版块</h3>
          <div class="advn-one">
            <div class="ad-mian">
              <div class="ad-left">
                <p>1</p>
              </div>
              <div class="ad-right">
                <h4><a @click="show('single')">视频课程</a></h4>
                <p>
                  视频课程涵盖明日科技的所有课程，针对不同用户的学习需求，用户可以根据自身情况选择适合
                  自己的学习方式。视频课程包括体系课程、实战课程等，体系课程可以让用户的学习更具有系统
                  性，同时能根据课程的周期，更有效的提高学习效率，优化学习效果。实战课程可以让用户通过
                  实例、项目、模块等实训来练习，让学习更有效。
                </p>
              </div>
              <div class="clearfix"></div>
            </div>
            <div class="ad-mian">
              <div class="ad-left">
                <p>2</p>
              </div>
              <div class="ad-right">
                <h4><a @click="show('single')">读书</a></h4>
                <p>读书是明日科技新上线的模块，也是明日科技多年编程项目、编程经验的积累，
                   适应各种用户学习，更有不断更新的电子书资源将陆续上线。   </p>
```

```html
            </div>
            <div class="clearfix"></div>
          </div>
          <div class="ad-mian">
            <div class="ad-left">
              <p>3</p>
            </div>
            <div class="ad-right">
              <h4><a @click="show('single')">社区</a></h4>
              <p>社区是明日科技用户交流的重要部分，可以进行技术讨论、下载相关资源以及灌水交流等。
                互动交流时，可以上传图片、附件、投票支持或投票反对，更可以@好友一起玩耍~</p>
            </div>
            <div class="clearfix"></div>
          </div>
        </div>
        <div class="col-md-6 advantages-left ">
          <h3>特色功能</h3>
          <div class="advn-two">
            <h4><a @click="show('single')"> 在线编辑器</a></h4>
            <p>在线编辑器为实践练习而专门设计的。在这里用户不需要搭建甚木开发环境，就可以在线做练习，
              使用户快速的掌握和运用知识。同时提供了在线问答、资料下载、名师点播、实例讲解等多种实用功能。</p>
            <ul>
              <li><a href="#"><i class="glyphicon glyphicon-ok"></i>边学习边提问</a></li>
              <li><a href="#"><i class="glyphicon glyphicon-ok"></i>3 天完成一个实用项目</a></li>
              <li><a href="#"><i class="glyphicon glyphicon-ok"></i>客服电话在线答疑</a></li>
              <li><a href="#"><i class="glyphicon glyphicon-ok"></i>移动端、PC 端同步网路学习</a></li>
            </ul>
          </div>
        </div>
        <div class="clearfix"></div>
      </div>
    <!--advantages-->
    <!--team-->
    <div class="team-us">
      <div class="team-top ">
        <h3>我们的团队</h3>
      </div>
      <div class="team-bottom">
        <ul class="ch-grid">
          <li v-for="(staff,index) in team" :key="staff">
            <div class="ch-item">
              <div class="ch-info-wrap">
                <div class="ch-info">
                  <div class="ch-info-front" :style="{backgroundImage:bgArr[index]}"></div>
                  <div class="ch-info-back">
                    <h3>{{ staff.name }}</h3>
                    <p>{{ staff.position }}</p>
                  </div>
                </div>
              </div>
            </div>
          </li>
        </ul>
      </div>
    </div>
  </div>
</template>
```

```
<script>
export default {
  name: "AboutHome",
  data: function (){
    return {
      team: [
        {name: 'Jonsen', position: '前端工程师'},
        {name: 'Livina', position: '网页设计师'},
        {name: 'Jefe', position: '后端工程师'}
      ],
      bgArr: [
        'url('+require('@/assets/images/team-1.jpg')+')',
        'url('+require('@/assets/images/team-2.jpg')+')',
        'url('+require('@/assets/images/team-3.jpg')+')'
      ]
    }
  },
  methods: {
    show: function (value){
      this.$router.push({name: value});
    }
  }
}
</script>
```

4.5.5 路由配置

下面给出该项目的路由配置,包括定义路由、创建路由对象、设置路由跳转后页面置顶,以及设置页面标题等功能。代码如下:

```
import { createRouter, createWebHistory } from 'vue-router'
import IndexHome from '@/views/index/IndexHome.vue'
import BlogHome from "@/views/blog/BlogHome";
import AboutHome from "@/views/about/AboutHome";
import SingleHome from "@/views/single/SingleHome";
const routes = [                                    //定义路由
  {
    path: '/',
    name: 'index',
    component: IndexHome,
    meta: {
      title: '主页'
    }
  },
  {
    path: '/about',
    name: 'about',
    component: AboutHome,
    meta: {
      title: '关于我们'
    }
  },
  {
    path: '/blog',
    name: 'blog',
    component: BlogHome,
    meta: {
      title: '博客列表页面'
    }
```

```
    },
    {
      path: '/single',
      name: 'single',
      component: SingleHome,
      meta: {
        title: '博客详情页面'
      }
    }
]
const router = createRouter({
  history: createWebHistory(process.env.BASE_URL),
  routes,
  //跳转页面后置顶
  scrollBehavior(to,from,savedPosition){
    if(savedPosition){
      return savedPosition;
    }else{
      return {top:0,left:0}
    }
  }
})
router.beforeEach((to, from, next) => {
  /*路由发生变化时修改页面 title */
  if (to.meta.title) {
    document.title = to.meta.title
  }
  next()
})
export default router
```

4.6 项目运行

通过前述步骤，我们已经设计并完成了"游戏公园博客"项目的开发。接下来，我们运行该项目，以检验我们的开发成果。首先打开命令提示符窗口，切换到项目所在目录，然后执行 npm run serve 命令运行该项目，如图 4.12 所示。

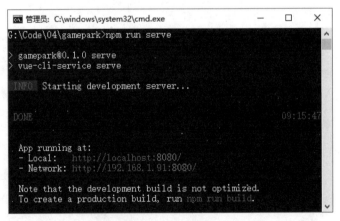

图 4.12 运行项目

在浏览器地址栏中输入 http://localhost:8080，然后按 Enter 键，即可成功运行该项目。游戏公园的主页效果如图 4.13 所示。

图 4.13　游戏公园主页效果

4.7　源码下载

本章虽然详细地讲解了如何编码实现"游戏公园博客"的各个功能，但给出的代码都是代码片段，而非完整的源代码。为了方便读者学习，本书提供了该项目的完整源代码，读者可以扫描右侧的二维码进行下载。

源码下载

第 5 章 电影易购 APP

——Vue CLI + Vue Router + Vuex + axios

在生活节奏日益加快的今天,很多人在闲暇之余喜欢通过看电影来放松身心。随着互联网的发展,使用电影购票 APP 购买电影票已经成为人们最主要的购票方式。本章将以电影购票为主题,使用 Vue CLI、Vue Router、Vuex 和 axios 等技术开发一个移动端应用——电影易购 APP。

项目微视频

本项目的核心功能及实现技术如下:

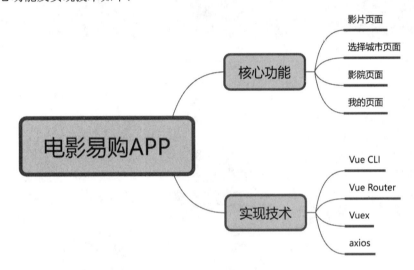

5.1 开发背景

随着社会的发展和生活水平的不断提高,人们对文化生活的需求不断增加,看电影已经成为人们休闲娱乐的主要方式之一。在周末、节假日或者工作日下班之余看一场电影,无疑是放松身心、丰富精神世界的极佳选择。据有关资料显示,近些年喜欢看电影的社会群体持续扩大,全国电影总票房屡创新高。购买电影票是人们看电影的首要步骤,电影院柜台采用人工售票的方式,不仅效率低、易出错,而且还耗费大量的人力。随着互联网技术的快速发展,网络购票业务快速地成长了起来,如今通过购票 APP 购买电影票已经取代了传统的电影院柜台购票方式。购票 APP 提供的在线购票功能不仅有助于电影院节省人力成本,还能显著提升用户的购票体验,并更好地发掘喜欢看电影的用户。本章将使用 Vue CLI、Vue Router、Vuex 和 axios 等技术开发一个电影购票的移动端应用——电影易购 APP。该应用的实现目标如下:

☑ 列表展示正在热映的影片。
☑ 列表展示即将上映的影片。
☑ 实现影片搜索功能。
☑ 实现按首字母检索城市和搜索城市的功能。

- ☑ 实现影院列表页面。
- ☑ 实现我的页面。

5.2 系统设计

5.2.1 开发环境

本项目的开发及运行环境如下：
- ☑ 操作系统：推荐 Windows 10、Windows 11 或更高版本，同时兼容 Windows 7（SP1）。
- ☑ 开发工具：WebStorm。
- ☑ 开发框架：Vue.js 3.0。

5.2.2 业务流程

电影易购 APP 由多个页面组成，包括正在热映和即将上映影片列表页面、搜索影片页面、城市列表页面、影院列表页面等。根据该项目的业务需求，我们设计如图 5.1 所示的业务流程图。

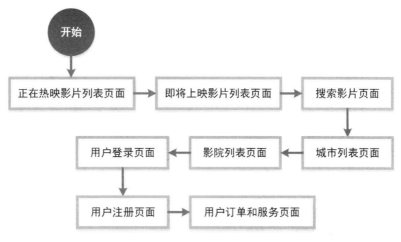

图 5.1 业务流程图

5.2.3 功能结构

本项目的功能结构已经在章首页中给出。其实现的具体功能如下：
- ☑ 正在热映影片列表页面：展示正在热映的影片信息，包括影片图片、影片名称、观众评分和影片主演等信息。
- ☑ 即将上映影片列表页面：展示即将上映的影片信息，包括影片图片、影片名称、想看人数、影片类型和影片主演等信息。
- ☑ 搜索影片页面：通过输入的影片名称关键字搜索影片。
- ☑ 城市列表页面：展示城市列表、根据首字母检索城市和搜索城市。
- ☑ 影院列表页面：展示所有影院信息，包括影院名称、影院地址、电影票最低价格，以及当前位置到影院的距离等信息。

- ☑ 注册页面：为用户提供注册功能。
- ☑ 登录页面：为用户提供登录功能。
- ☑ 用户订单与服务页面：用户登录后，展示用户订单和提供的服务选项。

5.3 技术准备

在开发电影易购 APP 时，我们主要使用 Vue CLI 脚手架工具、Vue Router、Vuex 和 axios 等技术。在该项目中：Vue CLI 脚手架工具用于创建项目；Vue Router 用于配置路由，实现页面之间的跳转；Vuex 用于保存登录账号、正在热映影片列表和即将上映影片列表；axios 用于发送请求，获取城市列表数据。这些知识在《Vue.js 从入门到精通》一书中有详细的讲解，对这些知识不太熟悉的读者，可以参考该书的相关章节。

5.4 创建项目

在设计电影易购 APP 各功能模块之前，需要使用 Vue CLI 创建项目，将项目名称设置为 ticket。在命令提示符窗口中输入以下命令：

```
vue create ticket
```

按 Enter 键，选择 Manually select features，如图 5.2 所示。

按 Enter 键后，选择 Router 和 Vuex 选项，如图 5.3 所示。

然后选择路由是否使用 history 模式，输入 y 表示使用 history 模式，如图 5.4 所示。

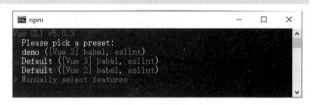

图 5.2　选择 Manually select features

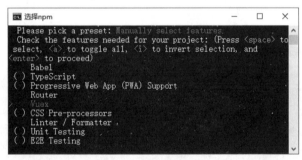

图 5.3　选择配置选项

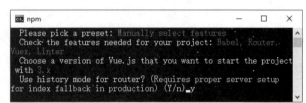

图 5.4　使用 history 模式

创建项目后，先整理项目的目录，然后在 assets 目录中创建 css 和 images 两个文件夹。其中：在 css 文件夹中创建 common.css 文件，该文件被用作项目的公共样式文件；在 images 文件夹中存储项目所需的影片图片。

5.5 公共组件设计

在开发项目时，编写公共组件可以减少重复代码的编写，这有助于提升代码的重用性和维护性。在设计电影易购 APP 时，所有页面都会用到两个公共组件：一个是页面头部组件 TheHeader.vue，另一个是底部导航栏组件 TabBar.vue。接下来，我们将详细介绍这两个组件。

5.5.1 头部组件设计

头部组件主要用于定义该页面的标题，其界面效果如图 5.5 所示。

头部组件 TheHeader.vue 的具体实现步骤如下。

（1）在<template>标签中添加一个<header>标签，在该标签中添加一个一级标题，并绑定父组件传递的 Prop 属性。代码如下：

图 5.5　头部组件界面效果

```
<template>
    <header id="header">
        <h1>{{title}}</h1>
    </header>
</template>
```

（2）在<script>标签中定义 Prop 属性，设置参数的数据类型和默认值，代码如下：

```
<script>
    export default {
        name: "TheHeader",
        props:{                              //定义 Prop 属性
            title:{
                type: String,
                default: '影片'
            }
        }
    }
</script>
```

（3）在<style>标签中，为头部组件中的元素设置样式，代码如下：

```
<style scoped>
    #header{
        width: 100%;                         /*设置宽度*/
        height: 50px;                        /*设置高度*/
        color: #ffffff;                      /*设置文字颜色*/
        background: #0097ef;                 /*设置背景颜色*/
        border-bottom:1px solid #0097ef;     /*设置下边框*/
        position: relative;                  /*设置相对定位*/
    }
    #header h1{
        font-size: 18px;                     /*设置文字大小*/
        text-align: center;                  /*设置文本水平居中显示*/
        line-height: 50px;                   /*设置行高*/
        font-weight: normal;                 /*设置文字粗细*/
    }
</style>
```

5.5.2 底部导航栏组件设计

底部导航栏组件主要用于页面之间的跳转，其界面效果如图 5.6 所示。

底部导航栏 TabBar.vue 的具体实现步骤如下。

（1）在<template>标签中添加一个 ul 列表，然后

图 5.6　底部导航栏组件界面效果

在该列表中使用<router-link>组件设置导航链接，并将<router-link>渲染为<li>标签。代码如下：

```html
<template>
    <div id="footer">
        <ul>
            <router-link to="/movie/onshow" custom v-slot="{navigate, isExactActive}">
                <li :class="[isExactActive && 'router-link-exact-active']" @click="navigate">
                    <i class="fa fa-film"></i>
                    <p>影片</p>
                </li>
            </router-link>
            <router-link to="/cinema" custom v-slot="{navigate, isExactActive}">
                <li :class="[isExactActive && 'router-link-exact-active']" @click="navigate">
                    <i class="fa fa-youtube-play"></i>
                    <p>影院</p>
                </li>
            </router-link>
            <router-link to="/my" custom v-slot="{navigate, isActive}">
                <li :class="[isActive && 'router-link-exact-active']" @click="navigate">
                    <i class="fa fa-user"></i>
                    <p>我的</p>
                </li>
            </router-link>
        </ul>
    </div>
</template>
```

> **说明**
>
> 如果想把<router-link>渲染为其他标签，并且希望将激活的路由链接能够高亮显示，可以在<router-link>中包含目标标签并使用 v-slot 来创建链接。navigate 表示触发导航的函数，而 isExactActive 用于指示需要应用精确激活的 CSS 类。

（2）在<style>标签中为底部导航栏组件中的元素设置样式，代码如下：

```css
<style scoped>
    #footer{
        width: 100%;                        /*设置宽度*/
        height: 50px;                       /*设置高度*/
        background: white;                  /*设置背景颜色*/
        border-top: 2px solid #ebe8e3;      /*设置上边框*/
        position: fixed;                    /*设置相对于浏览器进行定位*/
        left: 0;                            /*设置元素到浏览器左端距离*/
        bottom: 0;                          /*设置元素到浏览器底端距离*/
    }
    #footer ul{
        display: flex;                      /*设置弹性布局*/
        text-align: center;                 /*设置文本水平居中显示*/
        height: 50px;                       /*设置高度*/
        align-items: center;                /*设置每个元素在交叉轴上居中对齐*/
    }
    #footer ul li{
        flex: 1;                            /*设置元素宽度相等*/
        height: 40px;                       /*设置高度*/
    }
    #footer ul i{
        font-size: 20px;                    /*设置文字大小*/
    }
    #footer ul p{
        font-size: 12px;                    /*设置文字大小*/
```

```
            line-height: 18px;                        /*设置行高*/
        }
        .router-link-exact-active{
            color: #f03d37;                           /*设置激活文字颜色*/
        }
</style>
```

5.6 影片页面设计

与影片页面相关的组件包括正在热映影片组件、即将上映影片组件、影片搜索组件和影片页面组件。下面对这 4 个组件进行详细介绍。

5.6.1 正在热映影片组件设计

正在热映影片组件主要定义正在热映的影片信息，包括影片图片、影片名称、观众评分、影片主演、放映场次和"购票"按钮。该组件的页面效果如图 5.7 所示。

正在热映影片组件的实现过程如下：

（1）在 store/index.js 文件中定义正在热映影片列表，代码如下：

图 5.7 正在热映影片组件的页面效果

```
import { createStore } from 'vuex'
export default createStore({
    state: {
        hotmovies: [
            {
                img: require('@/assets/images/01.jpg'),
                name: '星际穿越',
                score: 9.1,
                stars: '马修·麦康纳 安妮·海瑟薇',
                times: '今天 50 家影院放映 800 场'
            },
            {
                img: require('@/assets/images/02.jpg'),
                name: '美女与野兽',
                score: 9.3,
                stars: '艾玛·沃森 丹·史蒂文斯',
                times: '今天 65 家影院放映 750 场'
            },
            {
                img: require('@/assets/images/03.jpg'),
                name: '阿凡达',
                score: 9.2,
                stars: '萨姆·沃辛顿 佐伊·索尔达娜 西格妮·韦弗',
                times: '今天 65 家影院放映 560 场'
            },
            {
                img: require('@/assets/images/04.jpg'),
                name: '疯狂原始人',
                score: 9.6,
                stars: '尼古拉斯·凯奇 玛·斯通',
                times: '今天 60 家影院放映 650 场'
            },
            {
```

```
            img: require('@/assets/images/05.jpg'),
            name: '美丽心灵',
            score: 9.5,
            stars: '罗素·克劳 艾德·哈里斯',
            times: '今天 56 家影院放映 520 场'
        },
        {
            img: require('@/assets/images/06.jpg'),
            name: '头号玩家',
            score: 9.6,
            stars: '泰尔·谢里丹 奥利维亚·库克 西蒙·佩吉',
            times: '今天 55 家影院放映 660 场'
        },
        {
            img: require('@/assets/images/07.jpg'),
            name: '侏罗纪世界 2',
            score: 9.3,
            stars: '克里斯·帕拉特 布莱丝·达拉斯·霍华德',
            times: '今天 70 家影院放映 600 场'
        },
        {
            img: require('@/assets/images/08.jpg'),
            name: '我在雨中等你',
            score: 8.6,
            stars: '米洛·文堤米利亚 凯文·科斯特纳 阿曼达·塞弗里德',
            times: '今天 50 家影院放映 300 场'
        }
    ]
  }
})
```

（2）在 components 文件夹下创建正在热映影片组件 OnShowing.vue。在<template>标签中定义一个 ul 列表，在该列表中对正在热映影片数组进行遍历，在遍历时输出影片图片、影片名称、观众评分和影片主演等信息。OnShowing.vue 组件的实现代码如下：

```
<template>
    <div class="movie_body">
        <ul>
            <li v-for="(item,index) in hotmovies" :key="index">
                <div class="pic_show"><img :src="item.img" alt=""></div>
                <div class="info_list">
                    <h2>{{ item.name }}</h2>
                    <p>观众评分 <span class="grade">{{ item.score }}</span></p>
                    <p>主演： {{ item.stars }}</p>
                    <p>{{ item.times }}</p>
                </div>
                <div class="btn_mall">
                    购票
                </div>
            </li>
        </ul>
    </div>
</template>
<script>
    import {mapState} from "vuex";                           //引入 mapState
    export default {
        name: "OnShowing",
        computed: {
            ...mapState([
                'hotmovies'
            ])
        }
```

```css
    }
</script>
<style scoped>
    .movie_body{
        overflow: auto;                              /*自动判断是否出现滚动条*/
    }
    .movie_body ul{
        margin: 0 12px;                              /*设置外边距*/
        overflow: hidden;                            /*设置溢出内容隐藏*/
    }
    .movie_body ul li{
        margin-top: 12px;                            /*设置上外边距*/
        display: flex;                               /*设置弹性布局*/
        align-items: center;                         /*设置每个元素在交叉轴上居中对齐*/
        border-bottom: 1px solid #e6e6e6;            /*设置下边框*/
        padding-bottom: 10px;                        /*设置下内边距*/
    }
    .movie_body .pic_show{
        width: 65px;                                 /*设置宽度*/
        height: 90px;                                /*设置高度*/
    }
    .movie_body .pic_show img{
        width: 100%;                                 /*设置宽度*/
    }
    .movie_body .info_list{
        margin-left:10px;                            /*设置左外边距*/
        flex: 1;                                     /*设置宽度为剩余空间*/
        position: relative;                          /*设置相对定位*/
    }
    .movie_body .info_list h2{
        font-size: 17px;                             /*设置文字大小*/
        line-height: 24px;                           /*设置行高*/
        width: 150px;                                /*设置宽度*/
        overflow: hidden;                            /*设置溢出内容隐藏*/
        white-space: nowrap;                         /*设置段落中的文本不进行换行*/
        text-overflow:ellipsis;                      /*溢出内容使用省略号*/
    }
    .movie_body .info_list p{
        font-size:13px;                              /*设置文字大小*/
        color: #666;                                 /*设置文字颜色*/
        line-height: 22px;                           /*设置行高*/
        width: 200px;                                /*设置宽度*/
        overflow: hidden;                            /*设置溢出内容隐藏*/
        white-space: nowrap;                         /*设置段落中的文本不进行换行*/
        text-overflow:ellipsis ;                     /*溢出内容使用省略号*/
    }
    .movie_body .info_list .grade{
        font-weight: 700;                            /*设置文字粗细*/
        color: #faaf00;                              /*设置文字颜色*/
        font-size: 15px;                             /*设置文字大小*/
    }
    .movie_body .info_list img{
        width: 50px;                                 /*设置宽度*/
        position: absolute;                          /*设置绝对定位*/
        right: 10px;                                 /*设置元素到父元素右侧距离*/
        top: 5px;                                    /*设置元素到父元素顶端距离*/
    }
    .movie_body .btn_mall{
        width: 47px;                                 /*设置宽度*/
        height: 27px;                                /*设置高度*/
        line-height: 28px;                           /*设置行高*/
        text-align: center;                          /*设置文本水平居中显示*/
```

```css
        background-color: #f03d37;              /*设置背景颜色*/
        color: #fff;                            /*设置文字颜色*/
        border-radius: 20px;                    /*设置圆角边框*/
        font-size: 12px;                        /*设置文字大小*/
        cursor: pointer;                        /*设置鼠标形状*/
    }
</style>
```

5.6.2 即将上映影片组件设计

即将上映影片组件主要定义即将上映的影片信息，包括影片图片、影片名称、想看人数、影片类型、影片主演和"预售"按钮。该组件的页面效果如图 5.8 所示。

即将上映影片组件的实现过程如下。

（1）在 store/index.js 文件中，在正在热映影片列表的下方添加即将上映影片列表，代码如下：

图 5.8　即将上映影片组件的页面效果

```js
upmovies: [
    {
        img: require('@/assets/images/09.jpg'),
        name: '在云端',
        number: 212465,
        type: '爱情、剧情',
        stars: '乔治·克鲁尼 维拉·法梅加 安娜·肯德里克'
    },
    {
        img: require('@/assets/images/10.jpg'),
        name: '奇异博士',
        number: 156765,
        type: '奇幻、动作、冒险',
        stars: '本尼迪克特·康伯巴奇 蒂尔达·斯文顿'
    },
    {
        img: require('@/assets/images/11.jpg'),
        name: '安娜·卡列尼娜',
        number: 128765,
        type: '剧情',
        stars: '凯拉·奈特莉 裘德·洛 亚伦·泰勒-约翰逊'
    },
    {
        img: require('@/assets/images/12.jpg'),
        name: '简爱',
        number: 66765,
        type: '剧情',
        stars: '迈克尔·法斯宾德 米娅·华希科沃斯卡'
    },
    {
        img: require('@/assets/images/13.jpg'),
        name: '阿凡达：水之道',
        number: 562375,
        type: '动作、科幻、冒险、奇幻',
        stars: '萨姆·沃辛顿 佐伊·索尔达娜 西格妮·韦弗'
    },
    {
        img: require('@/assets/images/14.jpg'),
        name: '蚁人',
        number: 236765,
        type: '动作、科幻、冒险、喜剧',
        stars: '保罗·路德 迈克尔·道格拉斯'
```

```
        },
        {
            img: require('@/assets/images/15.jpg'),
            name: '失控玩家',
            number: 138769,
            type: '科幻、动作、喜剧',
            stars: '瑞安·雷诺兹 朱迪·科默 乔·基瑞 塔伊加·维迪提'
        },
        {
            img: require('@/assets/images/16.jpg'),
            name: '勇敢者游戏：决战丛林',
            number: 236567,
            type: '冒险、动作、奇幻',
            stars: '道恩·强森 杰克·布莱克 凯文·哈特 凯伦·吉兰'
        }
    ]
```

（2）在 components 文件夹下创建即将上映影片组件 UpComing.vue。在<template>标签中定义一个 ul 列表，在该列表中对即将上映影片数组进行遍历，在遍历时输出影片图片、影片名称、想看人数和影片类型等信息。UpComing.vue 组件的实现代码如下：

```html
<template>
    <div class="movie_body">
        <ul>
            <li v-for="(item,index) in upmovies" :key="index">
                <div class="pic_show"><img :src="item.img" alt=""></div>
                <div class="info_list">
                    <h2>{{ item.name }}</h2>
                    <p><span class="person">{{ item.number }}</span>人想看</p>
                    <p>类型：{{ item.type }}</p>
                    <p>主演：{{ item.stars }}</p>
                </div>
                <div class="btn_pre">
                    预售
                </div>
            </li>
        </ul>
    </div>
</template>
<script>
    import {mapState} from "vuex";                      //引入 mapState
    export default {
        name: "UpComing",
        computed: {
            ...mapState([
                'upmovies'
            ])
        }
    }
</script>
<style scoped>
    .movie_body{
        overflow: auto;                                 /*自动判断是否出现滚动条*/
    }
    .movie_body ul{
        margin: 0 12px;                                 /*设置外边距*/
        overflow: hidden;                               /*设置溢出内容隐藏*/
    }
    .movie_body ul li{
        margin-top: 12px;                               /*设置上外边距*/
        display: flex;                                  /*设置弹性布局*/
        align-items: center;                            /*设置每个元素在交叉轴上居中对齐*/
```

```css
            border-bottom: 1px solid #e6e6e6;    /*设置下边框*/
            padding-bottom: 10px;                /*设置下内边距*/
        }
        .movie_body .pic_show{
            width: 64px;                         /*设置宽度*/
            height: 90px;                        /*设置高度*/
        }
        .movie_body .pic_show img{
            width: 100%;                         /*设置宽度*/
        }
        .movie_body .info_list{
            margin-left:10px;                    /*设置左外边距*/
            flex: 1;                             /*设置宽度为剩余空间*/
            position: relative;                  /*设置相对定位*/
        }
        .movie_body .info_list h2{
            font-size: 17px;                     /*设置文字大小*/
            line-height: 24px;                   /*设置行高*/
            width: 150px;                        /*设置宽度*/
            overflow: hidden;                    /*设置溢出内容隐藏*/
            white-space: nowrap;                 /*设置段落中的文本不进行换行*/
            text-overflow:ellipsis;              /*溢出内容使用省略号*/
        }
        .movie_body .info_list p{
            font-size:13px;                      /*设置文字大小*/
            color: #666;                         /*设置文字颜色*/
            line-height: 22px;                   /*设置行高*/
            width: 200px;                        /*设置宽度*/
            overflow: hidden;                    /*设置溢出内容隐藏*/
            white-space: nowrap;                 /*设置段落中的文本不进行换行*/
            text-overflow:ellipsis ;             /*溢出内容使用省略号*/
        }
        .movie_body .info_list img{
            width: 50px;                         /*设置宽度*/
            position: absolute;                  /*设置绝对定位*/
            right: 10px;                         /*设置元素到父元素右侧距离*/
            top: 5px;                            /*设置元素到父元素顶端距离*/
        }
        .movie_body .btn_pre{
            width: 47px;                         /*设置宽度*/
            height: 27px;                        /*设置高度*/
            line-height: 28px;                   /*设置行高*/
            text-align: center;                  /*设置文本水平居中显示*/
            background-color: #3c9fe6;           /*设置背景颜色*/
            color: #fff;                         /*设置文字颜色*/
            border-radius: 20px;                 /*设置圆角边框*/
            font-size: 12px;                     /*设置文字大小*/
            cursor: pointer;                     /*设置鼠标形状*/
        }
</style>
```

5.6.3 影片搜索组件设计

电影易购 APP 提供了搜索影片的功能。用户在搜索文本框中输入搜索关键字，下方会显示影片名称中包含搜索关键字的影片列表。该组件的页面效果如图 5.9 所示。

影片搜索组件的实现过程如下。

（1）在 components 文件夹下创建影片搜索组件

图 5.9 影片搜索组件的页面效果

SearchMovie.vue。在<template>标签中定义一个文本框和两个 ul 列表，使用 v-model 指令将该文本框和搜索关键字 keywords 进行绑定。在第一个 ul 列表中对搜索到的正在热映影片数组进行遍历，在第二个 ul 列表中对搜索到的即将上映影片数组进行遍历。代码如下：

```html
<template>
    <div class="search_body">
        <div class="search_input">
            <div class="search_input_wrapper">
                <i class="fa fa-search"></i>
                <input type="text" v-model="keywords">
            </div>
        </div>
        <div class="search_result">
            <ul>
                <li v-for="(item,index) in hotresult" :key="index">
                    <div class="pic_show"><img :src="item.img" alt=""></div>
                    <div class="info_list">
                        <h2>{{ item.name }}</h2>
                        <p>观众评分 <span class="grade">{{ item.score }}</span></p>
                        <p>主演：{{ item.stars }}</p>
                        <p>{{ item.times }}</p>
                    </div>
                    <div class="btn_mall">
                        购票
                    </div>
                </li>
            </ul>
            <ul>
                <li v-for="(item,index) in upresult" :key="index">
                    <div class="pic_show"><img :src="item.img" alt=""></div>
                    <div class="info_list">
                        <h2>{{ item.name }}</h2>
                        <p><span class="person">{{ item.number }}</span>人想看</p>
                        <p>类型：{{ item.type }}</p>
                        <p>主演：{{ item.stars }}</p>
                    </div>
                    <div class="btn_pre">
                        预售
                    </div>
                </li>
            </ul>
        </div>
    </div>
</template>
```

（2）在<script>标签中引入 mapState()辅助函数，实现组件中的计算属性和 store 中的 state 之间的映射。在 data 选项中定义搜索关键字 keywords，在 computed 选项中定义两个计算属性：hotresult 和 upresult。其中，hotresult 用于过滤正在热映影片数组中影片名称包含搜索关键字的影片，upresult 用于过滤即将上映影片数组中影片名称包含搜索关键字的影片。代码如下：

```html
<script>
    import {mapState} from "vuex";                    //引入 mapState
    export default {
        name: "SearchMovie",
        data(){
            return {
                keywords: ''                           //搜索关键字
            }
        },
        computed: {
```

```
      ...mapState([
        'hotmovies',
        'upmovies'
      ]),
      hotresult: function (){
        if(this.keywords === '') return;
        let t = this;
        //过滤正在热映影片数组
        return this.hotmovies.filter(function(item){
          return item.name.includes(t.keywords);
        })
      },
      //过滤即将上映影片数组
      upresult: function (){
        if(this.keywords === '') return;
        let t = this;
        return this.upmovies.filter(function(item){
          return item.name.includes(t.keywords);
        })
      }
    }
  }
</script>
```

（3）在<style>标签中为影片搜索组件中的元素设置样式，代码如下：

```
<style scoped>
  .search_body{
    overflow: auto;                                  /*自动判断是否出现滚动条*/
  }
  .search_body .search_input{
    padding: 8px 10px;                               /*设置内边距*/
    background-color: #f5f5f5;                       /*设置背景颜色*/
    border-bottom: 1px solid #e5e5e5;                /*设置下边框*/
  }
  .search_body .search_input_wrapper{
    padding: 0 10px;                                 /*设置内边距*/
    border: 1px solid #e6e6e6;                       /*设置边框*/
    border-radius: 5px;                              /*设置圆角边框*/
    background-color: #fff;                          /*设置背景颜色*/
    display: flex;                                   /*设置弹性布局*/
  }
  .search_body .search_input_wrapper i{
    font-size: 16px;                                 /*设置文字大小*/
    padding: 4px 0;                                  /*设置内边距*/
  }
  .search_body .search_input_wrapper input{
    border: none;                                    /*设置无边框*/
    font-size: 13px;                                 /*设置文字大小*/
    color: #333;                                     /*设置文字颜色*/
    padding: 4px;                                    /*设置内边距*/
    outline: none;                                   /*设置元素无轮廓*/
  }
  .search_body .search_result li{
    border-bottom: 1px #c9c9c9 dashed;                /*设置下边框*/
    padding: 10px 15px;                              /*设置内边距*/
    box-sizing: border-box;                          /*宽度和高度包括内容、内边距和边框*/
    display: flex;                                   /*设置弹性布局*/
    align-items: center;                             /*设置每个元素在交叉轴上居中对齐*/
  }
  .search_result .pic_show{
    width: 65px;                                     /*设置宽度*/
```

```css
        height: 90px;                                    /*设置高度*/
}
.search_result .pic_show img{
        width: 100%;                                     /*设置宽度*/
}
.search_result .info_list{
        margin-left:10px;                                /*设置左外边距*/
        flex: 1;                                         /*设置宽度为剩余空间*/
        position: relative;                              /*设置相对定位*/
}
.search_result .info_list h2, .search_result .info_list p{
        overflow: hidden;                                /*设置溢出内容隐藏*/
        white-space: nowrap;                             /*设置段落中的文本不进行换行*/
        text-overflow:ellipsis;                          /*溢出内容使用省略号*/
}
.search_result .info_list h2{
        font-size: 17px;                                 /*设置文字大小*/
        line-height: 24px;                               /*设置行高*/
        width: 150px;                                    /*设置宽度*/
}
.search_result .info_list p{
        font-size:13px;                                  /*设置文字大小*/
        color: #666;                                     /*设置文字颜色*/
        line-height: 22px;                               /*设置行高*/
        width: 200px;                                    /*设置宽度*/
}
.search_result .info_list .grade{
        font-weight: 700;                                /*设置文字粗细*/
        color: #faaf00;                                  /*设置文字颜色*/
        font-size: 15px;                                 /*设置文字大小*/
}
.search_result .info_list img{
        width: 50px;                                     /*设置宽度*/
        position: absolute;                              /*设置绝对定位*/
        right: 10px;                                     /*设置元素到父元素右端的距离*/
        top: 5px;                                        /*设置元素到父元素顶端的距离*/
}
.search_result .btn_mall, .search_result .btn_pre{
        width: 47px;                                     /*设置宽度*/
        height: 27px;                                    /*设置高度*/
        line-height: 28px;                               /*设置行高*/
        text-align: center;                              /*设置文本水平居中显示*/
        color: #fff;                                     /*设置文字颜色*/
        border-radius: 20px;                             /*设置圆角边框*/
        font-size: 12px;                                 /*设置文字大小*/
        cursor: pointer;                                 /*设置鼠标形状*/
}
.search_result .btn_mall{
        background-color: #f03d37;                       /*设置背景颜色*/
}
.search_result .btn_pre{
        background-color: #3c9fe6;                       /*设置背景颜色*/
}
</style>
```

5.6.4 影片页面组件设计

影片页面顶部有 4 个导航选项，分别对应选择城市页面、正在热映影片、即将上映影片和影片搜索组件。影片页面的效果如图 5.10 所示。

图 5.10　影片页面效果

影片页面组件的实现过程如下。

（1）在 views 文件夹下创建影片页面组件 MovieList.vue。在<template>标签中分别调用头部组件、定义<div>标签和调用底部导航栏组件。在<div>标签中，使用<router-link>定义导航选项，使用<router-view>渲染二级路由。代码如下：

```
<template>
    <div id="main">
        <TheHeader title="影片"></TheHeader>
        <div id="content">
            <div class="movie_menu">
                <router-link to="/city" custom v-slot="{navigate, isExactActive}">
                    <div :class="[isExactActive && 'router-link-exact-active']" class="city_name" @click="navigate">
                        <span>长春 </span><i class="fa fa-caret-down"></i>
                    </div>
                </router-link>
                <div class="hot_switch">
                    <router-link to="/movie/onshow" custom v-slot="{navigate, isExactActive}">
                        <div :class="[isExactActive && 'router-link-exact-active']" class="hot_item active" @click="navigate">
                            正在热映
                        </div>
                    </router-link>
                    <router-link to="/movie/coming" custom v-slot="{navigate, isExactActive}">
                        <div :class="[isExactActive && 'router-link-exact-active']" class="hot_item" @click="navigate">
                            即将上映
                        </div>
                    </router-link>
                </div>
                <router-link to="/movie/search" custom v-slot="{navigate, isExactActive}">
```

```
                <div :class="[isExactActive && 'router-link-exact-active']" class="search_entry" @click="navigate">
                    <i class="fa fa-search"></i>
                </div>
            </router-link>
        </div>
        <!--二级路由渲染-->
        <keep-alive>
            <router-view></router-view>
        </keep-alive>
    </div>
    <TabBar></TabBar>
</div>
</template>
```

（2）在<script>标签中分别引入头部组件和底部导航栏组件，在 components 选项中注册这两个组件。代码如下：

```
<script>
    import TheHeader from '../components/TheHeader';
    import TabBar from '../components/TabBar';
    export default {
        name:'MovieList',
        components:{
            TheHeader,
            TabBar
        }
    }
</script>
```

（3）在<style>标签中为影片页面组件中的元素设置样式，代码如下：

```
<style scoped>
    #content .movie_menu{
        width: 100%;                                /*设置宽度*/
        height: 45px;                               /*设置高度*/
        border-bottom: 1px solid #e6e6e6;           /*设置下边框*/
        display: flex;                              /*设置弹性布局*/
        justify-content: space-between;             /*设置两端对齐*/
    }
    .movie_menu .city_name{
        margin-left: 20px;                          /*设置左外边距*/
        height: 100%;                               /*设置高度*/
        line-height: 45px;                          /*设置行高*/
    }
    .router-link-exact-active{
        color: #ef4238;                             /*设置文字颜色*/
        border-bottom: 2px solid #ef4238;           /*设置下边框*/
        box-sizing: border-box;                     /*宽度和高度包括内容、内边距和边框*/
    }
    .movie_menu .hot_switch{
        display: flex;                              /*设置弹性布局*/
        height: 100%;                               /*设置高度*/
        line-height: 45px;                          /*设置行高*/
    }
    .movie_menu .hot_item{
        font-size: 15px;                            /*设置文字大小*/
        color: #666;                                /*设置文字颜色*/
        width: 80px;                                /*设置宽度*/
        text-align: center;                         /*设置文本水平居中显示*/
        margin: 0 12px;                             /*设置外边距*/
        font-weight: 700;                           /*设置文字粗细*/
```

```
    }
    .movie_menu .search_entry{
        margin-right: 20px;                          /*设置右外边距*/
        height: 100%;                                /*设置高度*/
        line-height: 45px;                           /*设置行高*/
    }
    .movie_menu .search_entry i{
        font-size: 24px;                             /*设置文字大小*/
        color: red;                                  /*设置文字颜色*/
    }
</style>
```

5.7 选择城市页面设计

选择城市页面主要包括搜索框、展示当前定位城市、热门城市、26 个大写字母和按照大写字母分类的城市列表。选择城市页面的效果如图 5.11 所示。用户单击任意一个大写字母，页面将展示该字母对应的城市列表，如图 5.12 所示。此外，用户在搜索框中输入搜索关键字后，页面下方将显示城市名称中包含搜索关键字的城市列表，如图 5.13 所示。

图 5.11　选择城市页面效果　　图 5.12　按字母分类的城市列表　　图 5.13　搜索结果中的城市列表

说明

在该项目中只是展示了当前定位的城市，而并未实现真正的定位功能。

选择城市页面的实现过程如下。

（1）在 public 文件夹下创建 city.json 文件，将选择城市页面中的城市列表数据保存在该文件中。代码如下：

```json
[
    {
        "name": "A",
        "citylist": ["阿坝","阿拉善","阿里","安康","安庆","鞍山","安顺","安阳"]
    },
    {
        "name": "B",
        "citylist": ["北京","白银","保定","宝鸡","保山","包头","巴中","北海","蚌埠","本溪","毕节","滨州","百色","亳州"]
    },
    {
        "name": "C",
        "citylist": ["重庆","成都","长沙","长春","沧州","常德","昌都","长治","常州","巢湖","潮州","承德","郴州","赤峰","池州","崇左","楚雄","滁州","朝阳"]
    },
    {
        "name": "D",
        "citylist": ["大连","东莞","大理","丹东","大庆","大同","大兴安岭","德宏","德阳","德州","定西","迪庆","东营"]
    },
    {
        "name": "E",
        "citylist": ["鄂尔多斯","恩施","鄂州"]
    },
    {
        "name": "F",
        "citylist": ["福州","防城港","佛山","抚顺","抚州","阜新","阜阳"]
    },
    {
        "name": "G",
        "citylist": ["广州","桂林","贵阳","甘南","赣州","甘孜","广安","广元","贵港","果洛"]
    },
    {
        "name": "H",
        "citylist": ["杭州","哈尔滨","合肥","海口","呼和浩特","海北","海东","海南","海西","邯郸","汉中","鹤壁","河池","鹤岗","黑河","衡水","衡阳","河源","贺州","红河","淮安","淮北","怀化","淮南","黄冈","黄南","黄山","黄石","惠州","葫芦岛","呼伦贝尔","湖州","菏泽"]
    },
    {
        "name": "J",
        "citylist": ["济南","佳木斯","吉安","江门","焦作","嘉兴","嘉峪关","揭阳","吉林","金昌","晋城","景德镇","荆门","荆州","金华","济宁","晋中","锦州","九江","酒泉"]
    },
    {
        "name": "K",
        "citylist": ["昆明","开封"]
    },
    {
        "name": "L",
        "citylist": ["兰州","来宾","莱芜","廊坊","乐山","凉山","连云港","聊城","辽阳","辽源","丽江","临沧","临汾","临夏","临沂","林芝","丽水","六安","六盘水","柳州","陇南","龙岩","娄底","漯河","洛阳","泸州","吕梁"]
    },
    {
        "name": "M",
        "citylist": ["马鞍山","茂名","眉山","梅州","绵阳","牡丹江"]
    },
    {
        "name": "N",
        "citylist": ["南京","南昌","南宁","宁波","南充","南平","南通","南阳","那曲","内江","宁德","怒江"]
    },
    {
        "name": "P",
```

```
      "citylist": ["盘锦","攀枝花","平顶山","平凉","萍乡","莆田","濮阳"]
    },
    {
      "name": "Q",
      "citylist": ["青岛","黔东南","黔南","黔西南","庆阳","清远","秦皇岛","钦州","齐齐哈尔","泉州","曲靖","衢州"]
    },
    {
      "name": "R",
      "citylist": ["日喀则","日照"]
    },
    {
      "name": "S",
      "citylist": ["上海","深圳","苏州","沈阳","石家庄","三门峡","三明","三亚","商洛","商丘","上饶","山南","汕头","汕尾","韶关","绍兴","邵阳","十堰","朔州","四平","绥化","遂宁","随州","宿迁","宿州"]
    },
    {
      "name": "T",
      "citylist": ["天津","太原","泰安","泰州","台州","唐山","天水","铁岭","铜川","通化","通辽","铜陵","铜仁"]
    },
    {
      "name": "W",
      "citylist": ["武汉","乌鲁木齐","无锡","威海","潍坊","文山","温州","乌海","芜湖","乌兰察布","武威","梧州"]
    },
    {
      "name": "X",
      "citylist": ["厦门","西安","西宁","襄樊","湘潭","湘西","咸宁","咸阳","孝感","邢台","新乡","信阳","新余","忻州","西双版纳","宣城","许昌","徐州","锡林郭勒","兴安"]
    },
    {
      "name": "Y",
      "citylist": ["银川","雅安","延安","延边","盐城","阳江","阳泉","扬州","烟台","宜宾","宜昌","宜春","营口","益阳","永州","岳阳","榆林","运城","云浮","玉树","玉溪","玉林"]
    },
    {
      "name": "Z",
      "citylist": ["漳州","湛江","肇庆","昭通","郑州","镇江","舟山","珠海","诸暨","驻马店","涿州","株洲","淄博","自贡","资阳"]
    }
]
```

（2）在 components 文件夹下创建选择城市页面组件 CityList.vue。在<template>标签中添加多个<div>标签，这些标签分别用于显示搜索框、搜索到的城市列表、当前定位城市、热门城市、所有城市列表和 26 个大写字母。当用户单击这些大写字母时，会调用 handleCity()方法。代码如下：

```
<template>
  <div id="main">
    <div class="header">
      <TheHeader title="选择城市"></TheHeader>
    </div>
    <div class="search_input">
      <div class="search_input_wrapper">
        <i class="fa fa-search"></i>
        <input type="text" v-model="keywords">
      </div>
    </div>
    <div class="city_body" v-show="keywords">
      <div class="city_list">
        <div class="city_sort">
          <div v-for="(item,index) in searchResult" :key="index">
            <h2>{{ item.name }}</h2>
            <ul>
              <li v-for="city in item.citylist" :key="city">{{ city }}</li>
```

```
            </ul>
          </div>
        </div>
      </div>
      <div class="city_body" v-show="!keywords">
        <div class="city_list">
          <div class="city_pos">
            <h2>定位城市</h2>
            <ul class="clearfix">
              <li>长春</li>
            </ul>
          </div>
          <div class="city_hot">
            <h2>热门城市</h2>
            <ul class="clearfix">
              <li v-for="item in hotcitys" :key="item">{{item}}</li>
            </ul>
          </div>
          <div class="city_sort">
            <div v-for="(item,index) in citylists" :key="index" :id="item.name">
              <h2>{{ item.name }}</h2>
              <ul>
                <li v-for="city in item.citylist" :key="city">{{ city }}</li>
              </ul>
            </div>
          </div>
        </div>
        <div class="city_index">
          <ul>
            <li v-for="letter in letters" :key="letter" @click="handleCity">{{ letter }}</li>
          </ul>
        </div>
      </div>
      <TabBar></TabBar>
    </div>
</template>
```

（3）在<script>标签中首先引入 axios、头部组件和底部导航栏组件，然后分别定义数据、注册的组件、mounted 钩子函数、方法、计算属性和监听属性。在 mounted 钩子函数中调用 setPosition()方法，并使用 axios 发送 GET 请求来获取城市列表数据。setPosition()方法用于设置搜索框、"定位城市"所在元素和"热门城市"所在元素到顶部的距离。handleCity()方法用于快速定位到指定大写字母对应的城市列表。searchResult 计算属性用于从城市数组中筛选出包含搜索关键字的城市。在 watch 选项中对 keywords 属性进行监听，当搜索关键字发生改变时设置搜索结果显示区域到顶部的距离。代码如下：

```
<script>
    import axios from "axios";                                  //引入 axios
    import TheHeader from "@/components/TheHeader.vue";
    import TabBar from "@/components/TabBar.vue";
    export default {
      name: "CityList",
      data(){
        return {
          hotcitys: ['北京','上海','广州','深圳','武汉','天津'],    //热门城市列表
          citylists: [],                                         //所有城市列表
          letters: ['A', 'B', 'C', 'D', 'E', 'F', 'G', 'H', 'I', 'J', 'K', 'L', 'M', 'N', 'O', 'P', 'Q', 'R', 'S', 'T', 'U', 'V', 'W', 'X', 'Y', 'Z'],
          toTop: 0,                                              //热门城市到顶部距离
          keywords: ''                                           //搜索关键字
        }
      },
```

```js
    components: {
      TheHeader,
      TabBar
    },
    mounted() {
        this.setPosition();
        axios.get('/city.json').then(function(response){
            this.citylists = response.data;
        }.bind(this));
    },
    methods: {
        setPosition(){
            const header = document.querySelector('.header');
            const search_input = document.querySelector('.search_input');
            const city_pos = document.querySelector('.city_pos');
            const city_hot = document.querySelector('.city_hot');
            //设置搜索框到顶部距离
            search_input.style.top = header.offsetHeight + 'px';
            //设置"定位城市"所在元素到顶部距离
            city_pos.style.top = header.offsetHeight + search_input.offsetHeight + 'px';
            this.toTop = header.offsetHeight + search_input.offsetHeight + city_pos.offsetHeight;
            city_hot.style.marginTop = this.toTop + 'px';          //设置"热门城市"所在元素到顶部距离
        },
        handleCity(e){
            const letter = e.target.innerHTML;
            if(letter === 'I' || letter === 'O' || letter === 'U' || letter === 'V') return;
            //窗口滚动的位置
            window.scrollTo({top: document.getElementById(letter).offsetTop - this.toTop, behavior: 'smooth' });
        }
    },
    computed: {
        searchResult(){
            let t = this;
            let newcitylists = [];                                 //搜索结果城市列表
            this.citylists.forEach(function (item){
                let arr =   item.citylist.filter(function (ite){
                    return ite.includes(t.keywords);               //从城市数组中筛选出包含搜索关键字的城市
                })
                if(arr.length){
                    newcitylists.push({
                        'name': item.name,
                        'citylist': arr
                    });
                }
            })
            return newcitylists;
        }
    },
    watch: {
        keywords(value){
            if(value){
                const header = document.querySelector('.header');
                const search_input = document.querySelector('.search_input');
                const city_body = document.querySelector('.city_body');
                //设置搜索结果显示区域到顶部距离
                city_body.style.marginTop = header.offsetHeight + search_input.offsetHeight + 'px';
            }
        }
    }
}
</script>
```

> **说明**
> 在选择城市页面组件中,获取城市数据是通过 axios 发送 GET 请求实现的,因此我们需要在项目中安装 axios,并且把保存城市数据的 JSON 文件存储在 public 文件夹中,以确保能够成功获取数据。

(4)在<style>标签中为选择城市页面组件中的元素设置样式,代码如下:

```css
<style scoped>
    .header{
        position: fixed;                              /*设置相对于浏览器进行定位*/
        width: 100%;                                  /*设置宽度*/
    }
    .search_input{
        position: fixed;                              /*设置相对于浏览器进行定位*/
        width: 95%;                                   /*设置宽度*/
        padding: 8px 10px;                            /*设置内边距*/
        background-color: #f5f5f5;                    /*设置背景颜色*/
        border-bottom: 1px solid #e5e5e5;             /*设置下边框*/
    }
    .search_input_wrapper{
        padding: 0 10px;                              /*设置内边距*/
        border: 1px solid #e6e6e6;                    /*设置边框*/
        border-radius: 5px;                           /*设置圆角边框*/
        background-color: #fff;                       /*设置背景颜色*/
        display: flex;                                /*设置弹性布局*/
    }
    .search_input_wrapper i{
        font-size: 16px;                              /*设置文字大小*/
        padding: 4px 0;                               /*设置内边距*/
    }
    .search_input_wrapper input{
        border: none;                                 /*设置无边框*/
        font-size: 13px;                              /*设置文字大小*/
        color: #333;                                  /*设置文字颜色*/
        padding: 4px;                                 /*设置内边距*/
        outline: none;                                /*设置元素无轮廓*/
    }
    .city_body{
        margin-top: 5px;                              /*设置上外边距*/
        display: flex;                                /*设置弹性布局*/
        width: 100%;                                  /*设置宽度*/
    }
    .city_body .city_list{
        overflow: auto;                               /*自动判断是否出现滚动条*/
        width: 92%;                                   /*设置宽度*/
    }
    .city_body .city_list::-webkit-scrollbar{
        background-color: transparent;                /*设置背景颜色透明*/
        width: 0;                                     /*设置宽度*/
    }
    .city_body .city_pos{
        position: fixed;                              /*设置相对于浏览器进行定位*/
        width: 100%;                                  /*设置宽度*/
        background-color: white;                      /*设置背景颜色*/
        padding: 10px 0;                              /*设置内边距*/
    }
    .city_body .city_pos h2,.city_body .city_hot h2{
        padding-left: 15px;                           /*设置左内边距*/
        line-height: 30px;                            /*设置行高*/
        font-size: 14px;                              /*设置文字大小*/
```

```css
        font-weight:normal;                         /*设置文字粗细*/
}
.city_body .city_pos ul li,.city_body .city_hot ul li{
        float: left;                                /*设置左浮动*/
        background: #fff;                           /*设置背景颜色*/
        width: 29%;                                 /*设置宽度*/
        height: 33px;                               /*设置高度*/
        margin-top: 15px;                           /*设置上外边距*/
        margin-left: 3%;                            /*设置左外边距*/
        padding:0 4px;                              /*设置内边距*/
        border: 1px solid #e6e6e6;                  /*设置边框*/
        border-radius: 3px;                         /*设置圆角边框*/
        line-height: 33px;                          /*设置行高*/
        text-align: center;                         /*设置文本水平居中显示*/
        box-sizing: border-box;                     /*宽度和高度包括内容、内边距和边框*/
}
.city_body .city_sort div{
        margin-top: 20px;                           /*设置上外边距*/
}
.city_body .city_sort h2{
        padding-left: 15px;                         /*设置左内边距*/
        line-height: 30px;                          /*设置行高*/
        font-size: 14px;                            /*设置文字大小*/
        font-weight: normal;                        /*设置文字粗细*/
}
.city_body .city_sort ul li{
        line-height: 36px;                          /*设置行高*/
        padding-left: 15px;                         /*设置左内边距*/
        border-bottom: 1px solid #CCC;              /*设置下边框*/
}
.city_body .city_index{
        font-size: 12px;                            /*设置文字大小*/
        margin-top: 130px;                          /*设置上外边距*/
        width: 20px;                                /*设置宽度*/
        position: fixed;                            /*设置定位方式*/
        right: 0;                                   /*设置元素到浏览器右边界距离*/
        display: flex;                              /*设置弹性布局*/
        justify-content: center;                    /*设置元素在主轴上居中排列*/
        text-align: center;                         /*设置文本水平居中显示*/
}
</style>
```

> **说明**
> 将搜索框、"定位城市"所在元素和"热门城市"所在元素的定位属性设置为相对于浏览器进行定位，这样可以实现元素不随滚动条滚动的效果。

5.8 影院页面设计

影院页面主要展示了当前定位城市的所有影院列表，包括影院名称、影院地址、电影票最低价格，以及影院和当前位置的距离等信息。影院页面的效果如图5.14所示。

> **说明**
> 影院和当前位置的距离是模拟数据，而不是使用定位技术实现的真正距离。

电影易购 APP 第 5 章

图 5.14 影院页面效果

5.8.1 影院列表组件设计

在设计影院页面时,我们决定将影院列表单独定义成一个组件。为此,我们在 components 文件夹下创建影院列表组件 CinemaList.vue。其实现过程如下。

(1) 在<template>标签中定义一个 ul 列表,在该列表中对影院列表进行遍历,在遍历时分别输出影院名称、电影票最低价格、影院地址、当前位置到影院的距离和卡券等信息。代码如下:

```
<template>
    <div class="cinema_body">
        <ul>
            <li v-for="item in cinemaList" :key="item">
                <div>
                    <span>{{ item.cinemaName }}</span>
                    <span class="q"><span class="price"> {{ item.startPrice }}元</span> 起</span>
                </div>
                <div class="address">
                    <span>{{ item.address }}</span>
                    <span> {{ item.distance }} </span>
                </div>
                <div class="card">
                    <div v-for="ite in item.cardList" :key="ite">{{ ite }}</div>
                </div>
            </li>
        </ul>
    </div>
</template>
```

（2）在<script>标签中定义影院列表，代码如下：

```
<script>
    export default {
        name: "CinemaList",
     data(){
        return {
            cinemaList: [                                              //影院列表
                {
                    cinemaName: '万达影城（欧亚卖场店）',                //影院名称
                    startPrice: 39,                                    //影票最低价格
                    address: '朝阳区开运街5178号欧亚卖场20号门4层',     //影院地址
                    distance: '800m',                                  //当前位置到影院的距离
                    cardList: ['小吃','影城卡']                         //卡券数组
                },
                {
                    cinemaName: '中影盈日国际影城（欧亚汇集店）',
                    startPrice: 33,
                    address: '朝阳区飞跃路2566号欧亚汇集4楼',
                    distance: '1.6km',
                    cardList: ['小吃','影城卡','券包']
                },
                {
                    cinemaName: '橙天NEX未来影城（钻石活力汇店）',
                    startPrice: 33,
                    address: '朝阳区重庆路1388号钻石活力汇5层',
                    distance: '2.1km',
                    cardList: ['小吃','券包']
                },
                {
                    cinemaName: '艾米1895电影街（长春鸿源店）',
                    startPrice: 39.9,
                    address: '朝阳区西安大路与同志街交汇新世纪鸿源4楼',
                    distance: '2.5km',
                    cardList: ['影城卡','券包']
                },
                {
                    cinemaName: '中影巨幕影城（欧亚商都店）',
                    startPrice: 23.9,
                    address: '朝阳区工农大路1128号欧亚商都10楼',
                    distance: '2.5km',
                    cardList: ['小吃','影城卡']
                }
            ]
        }
     }
    }
</script>
```

（3）在<style>标签中为影院列表组件中的元素设置样式，代码如下：

```
<style scoped>
    .cinema_body{
        overflow: auto;                                               /*自动判断是否出现滚动条*/
    }
    .cinema_body ul{
        padding: 20px;                                                /*设置内边距*/
    }
    .cinema_body li{
        border-bottom: 1px solid #e6e6e6;                             /*设置下边框*/
        margin-bottom: 20px;                                          /*设置下外边距*/
    }
```

```css
.cinema_body div{
    margin-bottom: 10px;                    /*设置下外边距*/
}
.cinema_body .q{
   float: right;                            /*设置右浮动*/
   font-size: 11px;                         /*设置文字大小*/
}
.cinema_body .price{
    font-size: 16px;                        /*设置文字大小*/
    color: #f03d37;                         /*设置文字颜色*/
}
.cinema_body .address{
    font-size: 13px;                        /*设置文字大小*/
    color:#666;                             /*设置文字颜色*/
}
.cinema_body .address span:nth-of-type(2){
    float: right;                           /*设置右浮动*/
}
.cinema_body .card{
    display: flex;                          /*设置弹性布局*/
}
.cinema_body .card div{
    font-size: 12px;                        /*设置文字大小*/
    margin-right: 2px;                      /*设置右外边距*/
    padding: 0 3px;                         /*设置内边距*/
    height: 15px;                           /*设置高度*/
    line-height: 15px;                      /*设置行高*/
    border-radius:2px;                      /*设置圆角边框*/
    color: #f60;                            /*设置文字颜色*/
    border:1px solid #f90;                  /*设置边框*/
}
</style>
```

5.8.2 影院页面组件设计

影院页面组件主要包括头部组件、影院列表组件和底部导航栏组件。在 views 文件夹下创建影院页面组件 TheCinema.vue，其实现过程如下：

（1）在<template>标签中定义<div>标签，在该标签中分别调用头部组件、影院列表组件和底部导航栏组件。代码如下：

```html
<template>
    <div id="main">
        <TheHeader title="影院"></TheHeader>
        <div id="content">
            <div class="cinema_menu">
                <div class="city_switch">
                    全城 <i class="fa fa-caret-down"></i>
                </div>
                <div class="city_switch">
                    品牌 <i class="fa fa-caret-down"></i>
                </div>
                <div class="city_switch">
                    筛选 <i class="fa fa-caret-down"></i>
                </div>
            </div>
            <CinemaList></CinemaList>
        </div>
        <TabBar></TabBar>
    </div>
```

```
</template>
```

（2）在<script>标签中分别引入头部组件、底部导航栏组件和影院列表组件，并在 components 选项中注册这 3 个组件。代码如下：

```
<script>
    import TheHeader from '../components/TheHeader';
    import TabBar from '../components/TabBar';
    import CinemaList from '../components/CinemaList';
    export default {
        name:'TheCinema',
        components:{
            TheHeader,
            TabBar,
            CinemaList
        }
    }
</script>
```

（3）在<style>标签中为影院页面组件中的元素设置样式，代码如下：

```
<style scoped>
#content .cinema_menu{
    width: 100%;                              /*设置宽度*/
    height: 45px;                             /*设置高度*/
    border-bottom: 1px solid #e6e6e6;         /*设置下边框*/
    display: flex;                            /*设置弹性布局*/
    justify-content: space-around;            /*设置每个元素两侧的间隔相等*/
    align-items: center;                      /*设置每个元素在交叉轴上居中对齐*/
    background: white;                        /*设置背景颜色*/
}
</style>
```

5.9 我的页面设计

与我的页面相关的组件包括用户登录组件、用户注册组件、用户订单与服务组件和我的页面组件。下面对这 4 个组件进行详细介绍。

5.9.1 用户登录组件设计

单击底部导航栏中的"我的"选项，如果用户未登录，系统将显示用户登录页面，以便用户进行登录操作。该页面的效果如图 5.15 所示。

用户登录组件的实现过程如下。

（1）在 store/index.js 文件中定义 state 和 mutation。其中，userName 表示用户账号，isLogin 表示用户是否已登录，login() 函数用于更改 state 中的状态。代码如下：

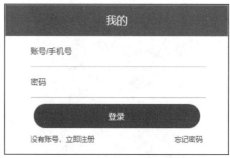

图 5.15 登录页面效果

```
import { createStore } from 'vuex'
export default createStore({
  state: {
    userName: null,            //账号
    isLogin: false,            //是否登录
  },
```

```js
  mutations: {
    login(state, userName){
      state.userName = userName;
      state.isLogin = true;
    }
  }
})
```

（2）在 components 文件夹下创建用户登录组件 UserLogin.vue。在<template>标签中定义用户登录表单，并将表单元素和定义的数据进行绑定。当用户单击"登录"按钮时，会调用 toHome()方法；当用户单击"没有账号，立即注册"超链接时，会调用 toRegister()方法，代码如下：

```html
<template>
    <div class="login_body">
        <form>
            <div>
                <input class="login_text" type="text" placeholder="账号" v-model="userName">
            </div>
            <div>
                <input class="login_text" type="password" placeholder="密码" v-model="userPwd">
            </div>
            <div class="login_btn">
                <input type="button" value="登录" @click="toHome">
            </div>
            <div class="login_link">
                <span @click="toRegister">没有账号，立即注册</span>
                <span>忘记密码</span>
            </div>
        </form>
    </div>
</template>
```

（3）在<script>标签中引入 mapState()和 mapMutations()辅助函数，以实现组件中的计算属性、方法和 store 中的 state、mutation 之间的映射。首先，定义 toRegister()方法，调用该方法会跳转到用户注册页面。然后，定义 toHome()方法，该方法首先会判断用户输入是否为空，接着验证用户输入的账号和密码是否正确。如果登录信息验证成功，则调用 login()方法，并跳转到用户订单和服务页面。代码如下：

```js
<script>
    import {mapState, mapMutations} from "vuex";              //引入 mapState 和 mapMutations
    export default {
        name: "UserLogin",
        data(){
            return {
                userName: null,                               //账号
                userPwd: null                                 //密码
            }
        },
        computed: {
            ...mapState([
                'userName',
                'isLogin'
            ])
        },
        methods: {
            ...mapMutations([
                'login'
            ]),
            toRegister(){
                this.$router.push({path: 'register'});        //跳转到用户注册页面
            },
```

```
        toHome(){
            let userName=this.userName;                              //获取账号
            let userPwd=this.userPwd;                                //获取密码
            if(userName === '' || userName === null){
                alert('请输入账号！');
                return false;
            }
            if(userPwd === '' || userPwd === null){
                alert('请输入密码！');
                return false;
            }
            if(userName !== 'mr' || userPwd !== 'mrsoft' ){
                alert('您输入的账号或密码有误！');
                return false;
            }else{
                alert('登录成功！');
                this.login(userName);
                this.$router.push({path: 'home'});                   //跳转到用户订单和服务页面
            }
        }
    },
    mounted() {
        if(this.isLogin){                                            //进入该页面先判断用户是否已登录
            this.$router.push({path: 'home'});                       //跳转到用户订单和服务页面
        }
    }
}
</script>
```

> **说明**
>
> 默认正确的账号为 mr，密码为 mrsoft。如果用户输入的账号或密码错误，系统将提示"您输入的账号或密码有误"；如果用户输入正确，系统将提示"登录成功"。

（4）在<style>标签中为用户登录组件中的元素设置样式，代码如下：

```
<style scoped>
    .login_body{
        width: 80%;                                    /*设置宽度*/
        margin: auto;                                  /*设置外边距*/
        overflow: hidden;                              /*设置溢出内容隐藏*/
    }
    .login_body .login_text{
        width: 100%;                                   /*设置宽度*/
        height: 50px;                                  /*设置高度*/
        border: none;                                  /*设置无边框*/
        border-bottom: 1px #ccc solid;                 /*设置下边框*/
        margin:0 5px;                                  /*设置外边距*/
        outline: none;                                 /*设置元素无轮廓*/
    }
    .login_body .login_btn{
        height: 40px;                                  /*设置高度*/
        margin: 10px;                                  /*设置外边距*/
    }
    .login_body .login_btn input{
        display: block;                                /*设置元素为块级元素*/
        width: 100%;                                   /*设置宽度*/
        height: 100%;                                  /*设置高度*/
        background: #e54847;                           /*设置背景颜色*/
        border-radius: 50px;                           /*设置圆角边框*/
        border: none;                                  /*设置无边框*/
```

```
            color: white;                                      /*设置文字颜色*/
        }
        .login_body .login_link{
            display: flex;                                     /*设置弹性布局*/
            justify-content: space-between;                    /*设置两端对齐*/
        }
        .login_body .login_link span{
            margin: 0 5px;                                     /*设置外边距*/
            font-size: 12px;                                   /*设置文字大小*/
            color:#e54847;                                     /*设置文字颜色*/
        }
</style>
```

5.9.2 用户注册组件设计

用户如果还未注册，则单击登录页面中的"没有账号，立即注册"超链接，将会跳转到用户注册页面。用户注册页面的效果如图 5.16 所示。

用户注册组件的实现过程如下。

（1）在 components 文件夹下创建用户注册组件 UserRegister.vue。在该组件的<template>标签中定义用户注册表单，并将表单元素和定义的数据进行绑定，当用户单击"注册"按钮时，会调用 register()方法，代码如下：

图 5.16 用户注册页面效果

```
<template>
    <div class="register_body">
        <form>
            <div>
                <input class="register_text" type="text" placeholder="账号" v-model="userName">
            </div>
            <div>
                <input class="register_text" type="password" placeholder="密码" v-model="userPwd">
            </div>
            <div>
                <input class="register_text" type="password" placeholder="确认密码" v-model="confirmPwd">
            </div>
            <div class="register_btn">
                <input type="button" value="注册" @click="register">
            </div>
        </form>
    </div>
</template>
```

（2）在<script>标签中定义数据，包括账号、密码和确认密码。定义 register()方法，该方法会判断用户输入是否为空，以及两次输入的密码是否一致。一旦用户注册成功，页面就会跳转到用户登录页面。代码如下：

```
<script>
    export default {
        name: "UserRegister",
        data(){
            return {
                userName: null,              //账号
                userPwd: null,               //密码
                confirmPwd: null             //确认密码
            }
```

```
        },
        methods: {
            register(){
                let userName=this.userName;                    //获取账号
                let userPwd=this.userPwd;                      //获取密码
                let confirmPwd=this.confirmPwd;                //获取确认密码
                //验证表单元素是否为空
                if(userName === '' || userName === null){
                    alert('账号不能为空！');
                    return false;
                }
                if(userPwd === '' || userPwd === null){
                    alert('密码不能为空！');
                    return false;
                }
                if(confirmPwd === '' || confirmPwd === null){
                    alert('确认密码不能为空！');
                    return false;
                }
                if(userPwd !== confirmPwd){
                    alert('两次密码不一致！');
                    return false;
                }
                alert('注册成功！');
                this.$router.push({path: 'login'});            //跳转到用户登录页面
            }
        }
    }
</script>
```

（3）在<style>标签中为用户注册组件中的元素设置样式，代码如下：

```
<style scoped>
    .register_body{
        width: 80%;                                            /*设置宽度*/
        margin: auto;                                          /*设置外边距*/
        overflow: hidden;                                      /*设置溢出内容隐藏*/
    }
    .register_body .register_text{
        width: 100%;                                           /*设置宽度*/
        height: 50px;                                          /*设置高度*/
        border: none;                                          /*设置无边框*/
        border-bottom: 1px #ccc solid;                         /*设置下边框*/
        margin:0 5px;                                          /*设置外边距*/
        outline: none;                                         /*设置元素无轮廓*/
    }
    .register_body .register_btn{
        height: 40px;                                          /*设置高度*/
        margin: 10px;                                          /*设置外边距*/
    }
    .register_body .register_btn input{
        display: block;                                        /*设置元素为块级元素*/
        width: 100%;                                           /*设置宽度*/
        height: 100%;                                          /*设置高度*/
        background: #e54847;                                   /*设置背景颜色*/
        border-radius: 50px;                                   /*设置圆角边框*/
        border: none;                                          /*设置无边框*/
        color: white;                                          /*设置文字颜色*/
    }
</style>
```

5.9.3 用户订单和服务组件设计

用户登录成功后,将会跳转到用户订单和服务页面。该页面主要展示用户订单和为用户提供的一些服务选项。该页面的效果如图 5.17 所示。

图 5.17 用户订单和服务页面效果

用户订单和服务组件的实现过程如下。

(1)在 components 文件夹下创建用户订单和服务组件 UserHome.vue。在该组件的<template>标签中定义多个<div>标签,并在这些<div>标签中分别添加用户头像、用户账号、用户订单选项列表和为用户提供的服务选项列表。代码如下:

```
<template>
    <div class="home_body">
        <div class="home_top">
            <img src="@/assets/images/head.gif">
            <p>{{userName}}</p>
        </div>
        <div class="home_order">
            <div>我的订单</div>
            <ul>
                <li>
                    <i class="fa fa-film"></i>
                    <p>电影</p>
                </li>
                <li>
                    <i class="fa fa-coffee"></i>
                    <p>小食周边</p>
                </li>
                <li>
```

```html
            <i class="fa fa-bullhorn"></i>
            <p>演出</p>
          </li>
        </ul>
      </div>
      <ul class="home_service">
        <li>
          <i class="fa fa-heart-o"></i>
          <p>想看的影片</p>
          <i class="fa fa-angle-right"></i>
        </li>
        <li>
          <i class="fa fa-play-circle-o"></i>
          <p>看过的影片</p>
          <i class="fa fa-angle-right"></i>
        </li>
        <li>
          <i class="fa fa-file-text-o"></i>
          <p>我的影评</p>
          <i class="fa fa-angle-right"></i>
        </li>
        <li>
          <i class="fa fa-envelope-o"></i>
          <p>我的消息</p>
          <i class="fa fa-angle-right"></i>
        </li>
        <li>
          <i class="fa fa-star"></i>
          <p>收藏的影院</p>
          <i class="fa fa-angle-right"></i>
        </li>
        <li>
          <i class="fa fa-question-circle-o"></i>
          <p>关于</p>
          <i class="fa fa-angle-right"></i>
        </li>
      </ul>
    </div>
</template>
```

（2）在<script>标签中引入 mapState()辅助函数，以实现组件中的计算属性和 store 中的 state 之间的映射。代码如下：

```html
<script>
    import {mapState} from "vuex";              //引入 mapState
    export default {
        name: "UserHome",
        computed: {
            ...mapState([
                'userName'
            ])
        }
    }
</script>
```

（3）在<style>标签中为用户订单和服务组件中的元素设置样式，代码如下：

```html
<style scoped>
    .home_body{
        width: 90%;                              /*设置宽度*/
        margin: auto;                            /*设置外边距*/
    }
```

```css
.home_body .home_top{
    display: flex;                          /*设置弹性布局*/
    align-items: center;                    /*设置每个元素在交叉轴上居中对齐*/
    height: 80px;                           /*设置高度*/
}
.home_body .home_top img{
    border-radius: 50%;                     /*设置图像显示为圆形*/
    width: 40px;                            /*设置宽度*/
    border: 1px solid #AAAAAA;              /*设置边框*/
}
.home_body .home_top p{
    margin-left: 10px;                      /*设置左外边距*/
}
.home_order{
    border-radius: 10px;                    /*设置圆角边框*/
    border: 1px solid #DDDDDD;              /*设置边框*/
    padding: 10px;                          /*设置内边距*/
}
.home_order div{
    height: 30px;                           /*设置高度*/
    line-height: 30px;                      /*设置行高*/
}
.home_order ul{
    display: flex;                          /*设置弹性布局*/
    text-align: center;                     /*设置文本水平居中显示*/
    height: 60px;                           /*设置高度*/
    align-items: center;                    /*设置每个元素在交叉轴上居中对齐*/
}
.home_order ul li{
    flex: 1;                                /*设置元素宽度相等*/
    height: 40px;                           /*设置高度*/
}
.home_order ul i{
    font-size: 20px;                        /*设置文字大小*/
}
.home_order ul p{
    font-size: 12px;                        /*设置文字大小*/
    line-height: 18px;                      /*设置行高*/
}
.home_service{
    margin-top: 30px;                       /*设置上外边距*/
    font-size: 14px;                        /*设置文字大小*/
}
.home_service li{
    display: flex;                          /*设置弹性布局*/
    height: 50px;                           /*设置高度*/
    align-items: center;                    /*设置每个元素在交叉轴上居中对齐*/
}
.home_service li:not(:last-child){
    border-bottom: 1px solid #999999;       /*设置下边框*/
}
.home_service li i:first-child{
    width: 30px;                            /*设置宽度*/
    margin-left: 3px;                       /*设置左外边距*/
    font-size: 16px;                        /*设置图标大小*/
}
.home_service li p{
    flex: 1;                                /*设置宽度为剩余空间*/
}
.home_service li i:last-child{
    width: 13px;                            /*设置宽度*/
    font-size: 16px;                        /*设置图标大小*/
```

5.9.4 我的页面组件设计

我的页面组件主要包括头部组件、二级路由渲染的组件和底部导航栏组件。在 views 文件夹下创建我的页面组件 MyIndex.vue。其实现过程如下：

（1）在<template>标签中分别调用头部组件、定义<div>标签和调用底部导航栏组件。在<div>标签中使用<router-view>渲染二级路由。代码如下：

```
<template>
    <div id="main">
        <TheHeader title="我的"></TheHeader>
        <div id="content">
            <router-view></router-view>
        </div>
        <TabBar></TabBar>
    </div>
</template>
```

（2）在<script>标签中分别引入头部组件和底部导航栏组件，在 components 选项中注册这两个组件。代码如下：

```
<script>
    import TheHeader from '../components/TheHeader';
    import TabBar from '../components/TabBar';
    export default {
        name: 'MyHome',
        components:{
            TheHeader,
            TabBar
        }
    }
</script>
```

5.10 路由配置

下面给出该项目的路由配置，包括引入页面组件、定义路由、创建路由对象，以及设置路由跳转后页面置顶。代码如下：

```
import { createRouter, createWebHistory } from 'vue-router'
import MovieList from "@/views/MovieList.vue";              //影片页面组件
import MyIndex from "@/views/MyIndex.vue";                  //我的页面组件
import TheCinema from "@/views/TheCinema.vue";              //影院页面组件
import CityList from "@/components/CityList.vue";           //选择城市页面组件
import OnShowing from "@/components/OnShowing.vue";         //正在热映页面组件
import UpComing from "@/components/UpComing.vue";           //即将上映页面组件
import SearchMovie from "@/components/SearchMovie.vue";     //影片搜索页面组件
import UserHome from "@/components/UserHome.vue";           //我的组件
import UserLogin from "@/components/UserLogin.vue";         //用户登录组件
import UserRegister from "@/components/UserRegister.vue";   //用户注册组件

const routes = [
    {
        path: '/movie',
        component: MovieList,
```

```
        children:[
            {
                path:'onshow',
                component: OnShowing
            },
            {
                path:'coming',
                component: UpComing
            },
            {
                path:'search',
                component: SearchMovie
            },
            {
                path:'',
                redirect: '/movie/onshow'
            }
        ]
    },
    {
        path:'/city',
        component: CityList
    },
    {
        path: '/cinema',
        component: TheCinema
    },
    {
        path: '/my',
        component: MyIndex,
        children: [
            {
                path: 'home',
                component: UserHome
            },
            {
                path: 'login',
                component: UserLogin
            },
            {
                path: 'register',
                component: UserRegister
            },
            {
                path: '',
                redirect: '/my/login'
            }
        ]
    },
    {
        path: '/:catchAll(.*)',
        redirect: '/movie/onshow'
    }
]
const router = createRouter({
    history: createWebHistory(process.env.BASE_URL),
    routes,
    //跳转页面后置顶
    scrollBehavior(to,from,savedPosition){
        if(savedPosition){
            return savedPosition;
        }else{
```

```
                return {top:0,left:0}
            }
        }
    })
export default router
```

5.11 项目运行

通过前述步骤，我们已经设计并完成了"电影易购 APP"项目的开发。接下来，我们运行该项目，以检验我们的开发成果。首先打开命令提示符窗口，切换到项目所在目录，执行 npm run serve 命令运行该项目，如图 5.18 所示。

在浏览器地址栏中输入 http://localhost:8080，然后按 Enter 键，即可成功运行该项目。默认情况下，页面会直接跳转至正在热映影片页面。该页面的效果如图 5.19 所示。

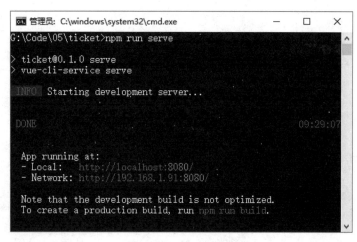

图 5.18 运行项目

图 5.19 正在热映影片页面效果

5.12 源码下载

本章虽然详细地讲解了如何编码实现"电影易购 APP"的各个功能，但给出的代码都是代码片段，而非完整的源代码。为了方便读者学习，本书提供了该项目的完整源代码，读者可以扫描右侧的二维码进行下载。

源码下载

第 6 章 淘贝电子商城

——Vue CLI + Vue Router + Vuex + localStorage

网络购物已经不再是什么新鲜事物，当今无论是企业，还是个人，都可以很方便地在网上交易商品，批发零售。比如，在淘宝上开网店，在微信上做微店，在抖音上开抖店等。本章将使用 Vue CLI、Vue Router、Vuex、localStorage 等技术设计并制作一个综合的电子商城项目——淘贝电子商城。

项目微视频

本项目的核心功能及实现技术如下：

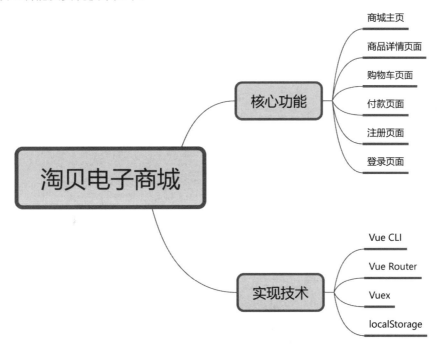

6.1 开发背景

随着互联网的快速发展，网上购物已经成为一种生活常态。网上购物和传统购物相比有着更加便捷的优势，加之现代化物流体系、网上支付体系的不断完善，使得消费者足不出户就可以买到自己需要的商品，享受安全便捷的购物过程。因此，越来越多的人开始在网上购物。在生活节奏越来越快的今天，人们需要有更多的休息时间，而网上购物可以给人们带来轻松和愉悦的购物体验。本章将使用 Vue CLI 、Vue Router、Vuex 和 localStorage 技术开发一个电子商城项目，该项目包含电子商务网站最基本的页面结构和功能，其实现目标如下：

☑ 通过网站主页展示促销活动和推荐商品。

- ☑ 通过商品详情页面展示商品的详细信息。
- ☑ 通过购物车页面展示用户选择的商品。
- ☑ 通过付款页面展示收货地址、订单信息和可以使用的优惠信息等内容，并可以提交订单，完成商品交易。
- ☑ 通过注册和登录页面实现用户的注册和登录。

6.2 系统设计

6.2.1 开发环境

本项目的开发及运行环境如下：
- ☑ 操作系统：推荐 Windows 10、Windows 11 或更高版本，同时兼容 Windows 7（SP1）。
- ☑ 开发工具：WebStorm。
- ☑ 开发框架：Vue.js 3.0。

6.2.2 业务流程

在设计淘贝电子商城时，用户如果未注册或已经注册但是未登录，则只能查看商城中的商品。用户只有在登录后才能进行选择商品、更改购物车中的商品数量、查看订单和提交订单等操作。

本项目的业务流程如图 6.1 所示。

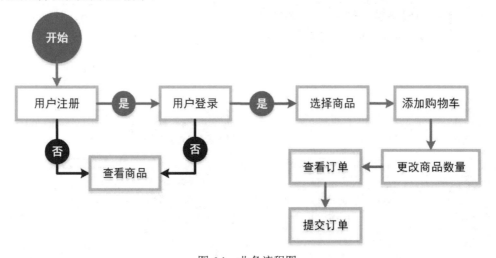

图 6.1 业务流程图

6.2.3 功能结构

本项目的功能结构已经在章首页中给出。作为一个电子商城的应用，本项目实现的具体功能如下：
- ☑ 商城主页：介绍重点推荐商品和促销商品等信息，具有分类导航功能，方便用户继续搜索商品。
- ☑ 商品详情页面：详细地展示具体某一种商品的信息，包括商品本身的介绍、购买商品后的评价、相似商品的推荐等内容。

- 购物车页面：详细记录已添加商品的价格和数量等内容，还可以进行更改商品数量和删除商品的操作。
- 付款页面：包含用户常用收货地址和订单信息等内容。
- 注册和登录页面：在用户注册或登录时进行表单信息提交和验证，如账号和密码不能为空、数字验证、邮箱验证等。

6.3 技术准备

在开发淘贝电子商城时，我们主要使用 Vue CLI 脚手架工具、Vue Router、Vuex 和 localStorage 等技术。Vue CLI 脚手架工具、Vue Router 和 Vuex 已经在前面的内容中做了简要介绍。这些知识在《Vue.js 从入门到精通》中有详细的讲解，对这些知识不太熟悉的读者，可以参考该书对应的内容。localStorage 是 HTML5 中的内容，下面将对 localStorage 进行必要介绍，以确保读者可以顺利完成本项目。

1．什么是 localStorage

HTML5 新增了 localStorage 特性，这个特性主要用于本地存储，解决了 cookie 存储空间不足的问题。localStorage 是持久化的本地存储，存储在其中的数据没有时间限制，永远不会过期，除非被手动删除。存储后的数据只要在相同的协议、相同的主机名、相同的端口下，就可以被读取或修改，从而实现跨页面共享同一份 localStorage 数据。

2．localStorage 常用属性和方法

- setItem(key,value)方法：设置 key 的值为 value。key 的类型是字符串类型。value 可以是包括字符串、布尔值、整数或者浮点数在内的任意 JavaScript 支持的类型。但是，最终数据是以字符串类型存储的。如果指定的 key 已经存在，那么新传入的数据会覆盖原来的数据。例如，使用 localStorage 对象保存键为 name、值为 mr 的数据，代码如下：

localStorage.setItem("name", "mr");

- length 属性：获得 localStorage 中保存数据项（item）的个数。例如，想要获取 localStorage 中存储的数据项的个数，可以使用下面的代码：

localStorage.length;

- key(index)方法：返回某个索引（从 0 开始计数）对应的 key，即 localStorage 中的第 index 个数据项的键。例如，想要获取 localStorage 中索引为 0 的数据项的键，可以使用下面的代码：

localStorage.key(0);

- getItem(key)方法：返回指定 key 的值，即存储的内容，类型为 String 字符串类型。如果传入的 key 不存在，那么该方法会返回 null，而不会抛出异常。例如，想要获取 localStorage 中 key 为 name 的数据项的值，可以使用下面的代码：

localStorage.getItem("name");

- removeItem(key)方法：从存储列表中删除指定的 key（包括对应的值）。例如，想要删除 localStorage 中 key 为 name 的数据项，可以使用下面的代码：

localStorage.removeItem("name");

☑ clear()方法：清空 localStorage 中的全部数据项。例如，想要清空 localStorage 中的数据项，可以使用下面的代码：

```
localStorage.clear();
```

6.4 主页的设计与实现

6.4.1 主页的设计

在越来越重视用户体验的今天，主页的设计非常关键。一个视觉效果出色且充满个性化的页面设计不仅能够给用户留下深刻印象，还能让用户流连忘返。为此，淘贝电子商城的主页精心设计了推荐商品和促销活动两个功能，旨在为用户呈现最新、最优质的商品和活动。主页的页面效果分别如图 6.2 和图 6.3 所示。

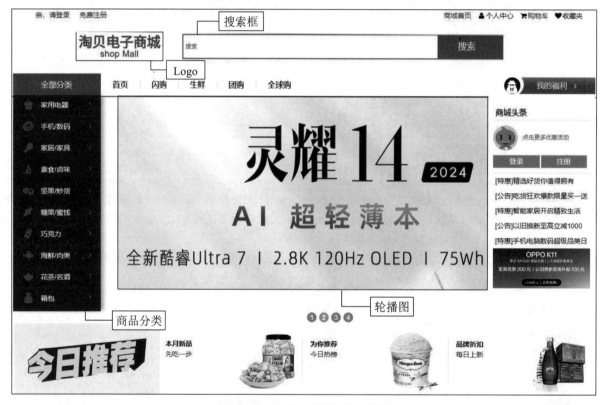

图 6.2 主页顶部区域的各个功能

6.4.2 顶部区和底部区功能的实现

根据由简到繁的原则，我们首先实现网站顶部区和底部区的功能。顶部区主要由网站的 Logo 图片、搜索框和导航菜单（登录、注册和商城首页等链接）组成，方便用户跳转到其他页面。底部区由制作公司和导航栏组成，链接到提供技术支持的官网。实现这些功能后的页面效果如图 6.4 所示。

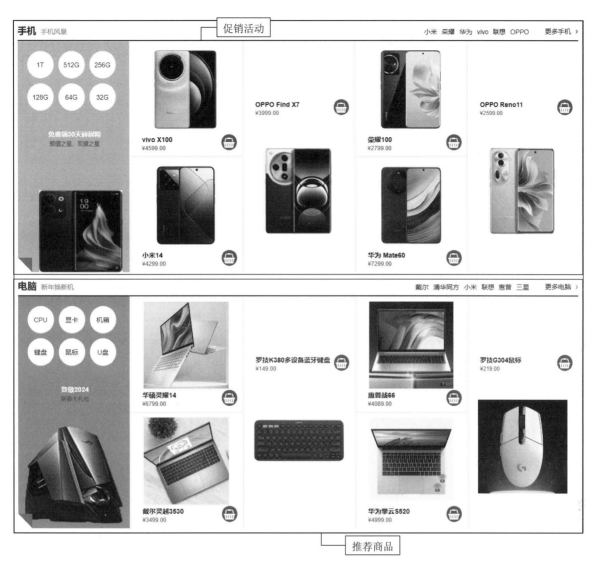

图 6.3 主页的促销活动区域和推荐商品区域

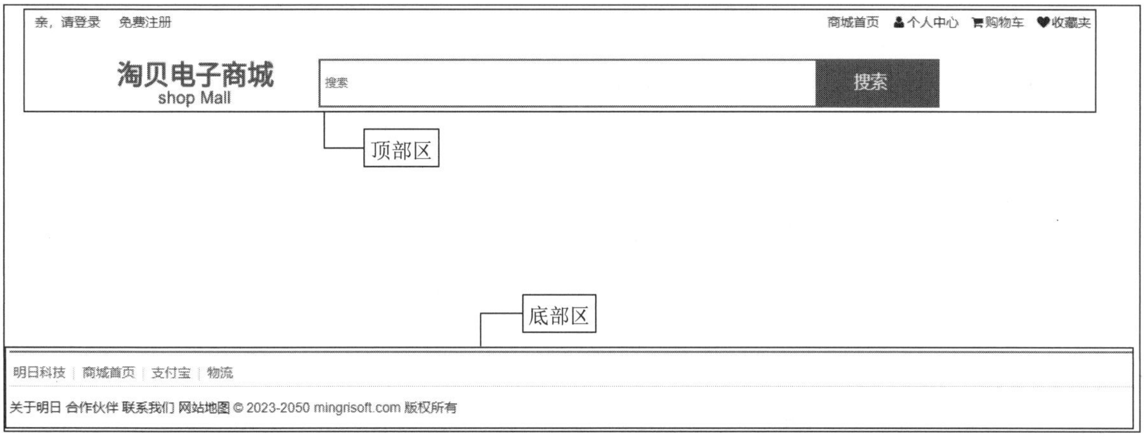

图 6.4 主页的顶部区和底部区

具体实现步骤如下。

（1）在 components 文件夹下新建 TheTop.vue 文件，用于实现顶部区的功能。在<template>标签中，定义导航菜单、网站的 Logo 图片和搜索框。在<script>标签中，判断用户的登录状态，以实现不同状态下页面的跳转。关键代码如下：

```html
<template>
  <div class="hmtop">
    <!--顶部导航条 -->
    <div class="mr-container header">
      <ul class="message-l">
        <div class="topMessage">
          <div class="menu-hd">
            <a @click="show('login')" target="_top" class="h" style="color: red" v-if="!isLogin">亲，请登录</a>
            <span v-else style="color: green">{{user}}，欢迎您 <a @click="logout" style="color: red">退出登录</a></span>
            <a @click="show('register')" target="_top" style="color: red; margin-left: 20px;">免费注册</a>
          </div>
        </div>
      </ul>
      <ul class="message-r">
        <div class="topMessage home">
          <div class="menu-hd"><a @click="show('home')" target="_top" class="h" style="color:red">商城首页</a></div>
        </div>
        <div class="topMessage my-shangcheng">
          <div class="menu-hd MyShangcheng">
            <a href="#" target="_top"><i class="mr-icon-user mr-icon-fw"></i>个人中心</a>
          </div>
        </div>
        <div class="topMessage mini-cart">
          <div class="menu-hd"><a id="mc-menu-hd" @click="show('shopcart')" target="_top">
            <i class="mr-icon-shopping-cart   mr-icon-fw" ></i><span style="color:red">购物车</span>
            <strong id="J_MiniCartNum" class="h" v-if="isLogin">{{length}}</strong>
          </a>
          </div>
        </div>
        <div class="topMessage favorite">
          <div class="menu-hd">
            <a href="#" target="_top"><i class="mr-icon-heart mr-icon-fw"></i><span>收藏夹</span></a>
          </div>
        </div>
      </ul>
    </div>
    <!--悬浮搜索框-->
    <div class="nav white">
      <div class="logo"><a @click="show('home')"><img src="@/assets/images/logo.png"/></a></div>
      <div class="logoBig">
        <li @click="show('home')"><img src="@/assets/images/logobig.png"/></li>
      </div>
      <div class="search-bar pr">
        <form>
          <input id="searchInput" name="index_none_header_sysc" type="text" placeholder="搜索" autocomplete="off">
          <input id="ai-topsearch" class="submit mr-btn" value="搜索" index="1" type="submit">
        </form>
      </div>
    </div>
    <div class="clear"></div>
  </div>
</template>
<script>
  import {mapState,mapGetters,mapActions} from 'vuex'     //引入辅助函数
  export default {
    name: 'TheTop',
```

```
    computed: {
        ...mapState([
            'user',                                     //将 this.user 映射为 this.$store.state.user
            'isLogin'                                   //将 this.isLogin 映射为 this.$store.state.isLogin
        ]),
        ...mapGetters([
            'length'                                    //将 this.length 映射为 this.$store.getters.length
        ])
    },
    methods: {
        show: function (value) {
            if(value == 'shopcart'){
                if(this.user == null){
                    alert('亲，请登录！');               //用户未登录
                    this.$router.push({name:'login'});  //跳转到登录页面
                    return false;
                }
            }
            this.$router.push({name:value});
        },
        ...mapActions([
            'logoutAction'                              //将 this.logoutAction()映射为 this.$store.dispatch('logoutAction')
        ]),
        logout: function () {
            if(confirm('确定退出登录吗？')){
                this.logoutAction();                    //执行退出登录操作
                this.$router.push({name:'home'});       //跳转到主页
            }else{
                return false;
            }
        }
    }
}
</script>
<style scoped lang="scss">
.logoBig li{
    cursor: pointer;                                    //定义鼠标指针形状
}
a{
    cursor: pointer;                                    //定义鼠标指针形状
}
</style>
```

（2）在 components 文件夹下新建 TheFooter.vue 文件，用于实现底部区的功能。在<template>标签中，首先通过<p>标签和<a>标签，实现底部的导航栏。然后使用<p>段落标签，显示关于明日、合作伙伴和联系我们等网站制作团队的相关信息。在<script>标签中，定义用于实现页面跳转的方法。代码如下：

```
<template>
    <div class="footer ">
        <div class="footer-hd ">
            <p>
                <a href="http://www.mingrisoft.com/" target="_blank">明日科技</a>
                <b>|</b>
                <a href="javascript:void(0)" @click="show">商城首页</a>
                <b>|</b>
                <a href="javascript:void(0)">支付宝</a>
                <b>|</b>
                <a href="javascript:void(0)">物流</a>
            </p>
        </div>
        <div class="footer-bd ">
            <p>
```

```
            <a href="http://www.mingrisoft.com/Index/ServiceCenter/aboutus.html" target="_blank">关于明日 </a>
            <a href="javascript:void(0)">合作伙伴 </a>
            <a href="javascript:void(0)">联系我们 </a>
            <a href="javascript:void(0)">网站地图 </a>
            <em>© 2023-2050 mingrisoft.com 版权所有</em>
        </p>
    </div>
  </div>
</template>
<script>
export default {
  methods: {
    show: function () {
      this.$router.push({name:'home'});               //跳转到主页
    }
  }
}
</script>
```

6.4.3 商品分类导航功能的实现

主页商品分类导航功能旨在将商品分门别类，便于用户检索和查找。用户将鼠标移到某一商品分类时，界面会继续弹出商品的子类别内容；用户将鼠标移出时，子类别内容会消失。因此，商品分类导航功能可以使商品信息更清晰易查，井井有条。实现后的页面效果如图 6.5 所示。

图 6.5　商品分类导航功能页面效果

具体实现步骤如下。

（1）在 views/index 文件夹下新建 IndexMenu.vue 文件。在<template>标签中，通过<li>标签显示商品分类信息。在<li>标签中，通过触发 mouseover 事件和 mouseout 事件执行相应的方法。关键代码如下：

```
<template>
    <div>
        <!--侧边导航 -->
        <div id="nav" class="navfull">
            <div class="area clearfix">
                <div class="category-content" id="guide_2">
                    <div class="category">
                        <ul class="category-list" id="js_climit_li">
                            <li class="appliance js_toggle relative" v-for="(v,i) in data" :key=
                                "i" @mouseover="mouseOver(i)" @mouseout="mouseOut(i)">
                                <div class="category-info">
```

```html
                    <h3 class="category-name b-category-name">
                        <i><img :src="v.url"></i>
                        <a class="ml-22" :title="v.bigtype">{{v.bigtype}}</a>
                    </h3>
                    <em>&gt;</em></div>
                    <div class="menu-item menu-in top" >
                        <div class="area-in">
                            <div class="area-bg">
                                <div class="menu-srot">
                                    <div class="sort-side">
                                        <dl class="dl-sort" v-for="v in v.smalltype" :key="v">
                                            <dt><span >{{v.name}}</span></dt>
                                            <dd v-for="v in v.goods" :key="v">
                                                <a href="javascript:void(0)"><span>{{v}}</span></a>
                                            </dd>
                                        </dl>
                                    </div>
                                </div>
                            </div>
                        </div>
                        <b class="arrow"></b>
                    </li>
                </ul>
            </div>
        </div>
    </div>
</div>
</template>
```

（2）在<script>标签中，编写鼠标移入和移出事件执行的方法。mouseOver()方法和 mouseOut()方法分别为鼠标移入和移出事件的方法，二者实现逻辑相似。以 mouseOver()方法为例，当鼠标移入<li>标签节点时，该方法会获取事件对象 obj，设置 obj 对象的样式，找到 obj 对象的子节点（子分类信息），将子节点内容显示到页面中。代码如下：

```
<script>
    import data from '@/assets/js/data.js';                    //导入数据
    export default {
        name: 'IndexMenu',
        data: function(){
            return {
                data: data
            }
        },
        methods: {
            mouseOver: function (i){
                var obj=document.getElementsByClassName('appliance')[i];
                obj.className="appliance js_toggle relative hover";   //设置当前事件对象样式
                var menu=obj.childNodes;                              //寻找该事件子节点（商品子类别）
                menu[1].style.display='block';                        //显示子节点
            },
            mouseOut: function (i){
                var obj=document.getElementsByClassName('appliance')[i];
                obj.className="appliance js_toggle relative";         //设置当前事件对象样式
                var menu=obj.childNodes;                              //寻找该事件子节点（商品子类别）
                menu[1].style.display='none';                         //隐藏子节点
            },
            show: function (value) {
                this.$router.push({name:value})
            }
```

      }
    }
</script>
```

### 6.4.4 轮播图功能的实现

轮播图功能根据固定的时间间隔，动态地显示或隐藏轮播图片，以此来引起用户的关注和注意。这些轮播图片一般都是系统推荐的最新商品内容。在主页中，我们通过应用过渡效果来实现图片的轮播。主页轮播图的页面效果如图 6.6 所示。

图 6.6　主页轮播图页面效果

具体实现步骤如下。

（1）在 views/index 文件夹下新建 IndexBanner.vue 文件。在<template>标签中使用 v-for 和<transition-group>组件实现列表过渡。在<li>标签中应用 v-for 指令定义 4 个数字轮播顺序节点。关键代码如下：

```
<template>
  <div class="banner">
    <div class="mr-slider mr-slider-default scoll" data-mr-flexslider id="demo-slider-0">
      <div id="box">
        <ul id="imagesUI" class="list">
          <transition-group name="fade" tag="div">
            <li v-for="(v,i) in banners" :key="v" v-show="(i+1)==index?true:false"><img :src="v"></li>
          </transition-group>
        </ul>
        <ul id="btnUI" class="count">
          <li v-for="num in 4" :key="num" @mouseover='change(num)' :class='{current:num==index}'>
            {{num}}
          </li>
        </ul>
      </div>
    </div>
    <div class="clear"></div>
  </div>
</template>
```

（2）在<script>标签中编写实现图片轮播的代码。在 mounted 钩子函数中，定义每经过 3 秒实现图片的轮换。在 change()方法中实现当鼠标移入数字按钮时，切换到对应的图片。关键代码如下：
```

```
<script>
  export default {
    name: 'IndexBanner',
    data : function(){
      return {
        banners : [                                    //广告图片数组
          require('@/assets/images/ad1.png'),
          require('@/assets/images/ad2.png'),
          require('@/assets/images/ad3.png'),
          require('@/assets/images/ad4.png')
        ],
        index : 1,                                     //图片的索引
        flag : true,
        timer : '',                                    //定时器ID
      }
    },
    methods : {
      next : function(){
        //下一张图片，图片索引为4时返回第一张
        this.index = this.index + 1 == 5 ? 1 : this.index + 1;
      },
      change : function(num){
        //当鼠标移入按钮时，切换到对应的图片
        if(this.flag){
          this.flag = false;
          //过1秒后，可以再次将鼠标移入按钮以切换图片
          setTimeout(()=>{
            this.flag = true;
          },1000);
          this.index = num;                            //切换为选中的图片
          clearTimeout(this.timer);                    //取消定时器
          //过3秒图片轮换
          this.timer = setInterval(this.next,3000);
        }
      }
    },
    mounted : function(){
      //过3秒图片轮换
      this.timer = setInterval(this.next,3000);
    }
  }
</script>
```

（3）在<style>标签中编写用于实现图片显示与隐藏过渡效果的类名。关键代码如下：

```
<style lang="scss" scoped>
/* 设置过渡属性 */
.fade-enter-active, .fade-leave-active{
  transition: all 1s;
}
.fade-enter-from, .fade-leave-to{
  opacity: 0;
}
</style>
```

6.4.5 商品推荐功能的实现

商品推荐功能是淘贝电子商城主要的商品促销形式，此功能可以动态显示推荐的商品信息，包括商品的缩略图、价格和打折信息等内容。商品推荐功能还能对众多商品信息进行精挑细选，从而提高商品的销售率。商品推荐功能的页面效果如图6.7所示。

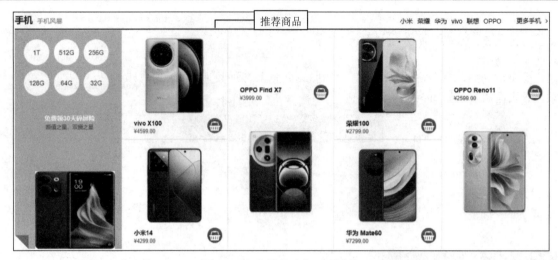

图 6.7 商品推荐功能页面效果

具体实现步骤如下。

（1）在 views/index 文件夹下新建 IndexPhone.vue 文件。在<template>标签中编写 HTML 的布局代码，使用 v-for 指令循环输出手机的品牌和核数。再通过<div>标签显示具体的商品内容，包括商品图片、名称和价格信息等。关键代码如下：

```
<template>
  <!--手机-->
  <div id="f1">
    <div class="mr-container ">
      <div class="shopTitle ">
        <h4>手机</h4>
        <h3>手机风暴</h3>
        <div class="today-brands ">
          <a href="javascript:void(0)" v-for="item in brands" :key="item">{{item}}</a>
        </div>
        <span class="more ">
          <a href="javascript:void(0)">更多手机<i class="mr-icon-angle-right" style="padding-left:10px ;"></i></a>
        </span>
      </div>
    </div>
    <div class="mr-g mr-g-fixed floodFive ">
      <div class="mr-u-sm-5 mr-u-md-3 text-one list">
        <div class="word">
          <a class="outer" href="javascript:void(0)" v-for="item in storage" :key="item">
            <span class="inner"><b class="text">{{item}}</b></span>
          </a>
        </div>
        <a href="javascript:void(0)">
          <img src="@/assets/images/tel.png" width="100px" height="170px"/>
          <div class="outer-con">
            <div class="title">
              免费领 30 天碎屏险
            </div>
            <div class="sub-title ">
              颜值之星，双摄之星
            </div>
          </div>
        </a>
        <div class="triangle-topright"></div>
      </div>
```

```html
<div class="mr-u-sm-7 mr-u-md-5 mr-u-lg-2 text-two">
    <div class="outer-con ">
        <div class="title ">
            vivo X100
        </div>
        <div class="sub-title ">
            ¥4599.00
        </div>
        <i class="mr-icon-shopping-basket mr-icon-md seprate"></i>
    </div>
    <a href="javascript:void(0)" @click="show"><img src="@/assets/images/phone1.jpg"/></a>
</div>
<!-- 省略部分代码 -->
</div>
<div class="clear "></div>
</div>
</template>
```

（2）在<script>标签中定义手机品牌数组和手机核数数组，定义当单击商品图片时执行的方法 show()，实现跳转到商品详情页面的功能。关键代码如下：

```
<script>
    export default {
        name: 'IndexPhone',
        data: function(){
            return {
                //手机品牌数组
                brands: ['小米','荣耀','华为','vivo','联想','OPPO'],
                //手机内存数组
                storage: ['1T','512G','256G','128G','64G','32G']
            }
        },
        methods: {
            show: function () {
                this.$router.push({name:'shopinfo'});      //跳转到商品详情页面
            }
        }
    }
</script>
```

> **说明**
> 当鼠标移入某具体的商品图片时，图片会呈现出一种偏移效果，这可以引起用户的注意并激发他们的兴趣。

6.5 商品详情页面的设计与实现

6.5.1 商品详情页面的设计

商品详情页面是商城主页的子页面。用户单击主页中的某一商品图片后，则进入商品详情页面。对于用户而言，商品详情页面是至关重要的功能页面。商品详情页面的界面和功能直接影响用户的购买意愿。为此，淘贝电子商城设计并实现了一系列的功能，包括商品图片放大镜效果、商品概要信息、宝贝详情和评价等功能模块，旨在帮助用户做出消费决策，从而提高商品销售量。商品详情页面的效果分别如图 6.8 和图 6.9 所示。

图 6.8　商品图片和概要信息

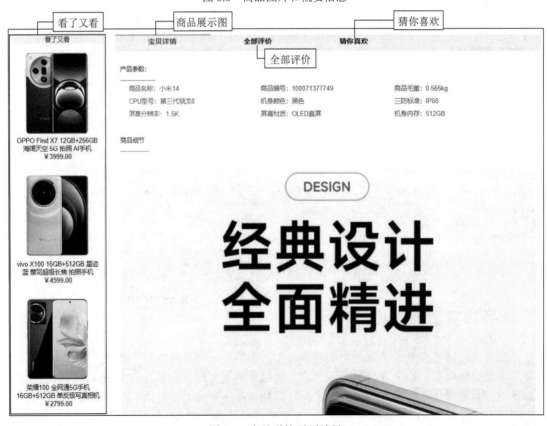

图 6.9　商品详情页面效果

6.5.2 图片放大镜效果的实现

商品展示图区域底部包含一个缩略图列表。当用户将鼠标指向某个缩略图时，上方会显示对应的商品图片。当用户将鼠标移入图片时，右侧会显示该图片对应区域的放大镜效果。图片放大镜效果如图 6.10 所示。

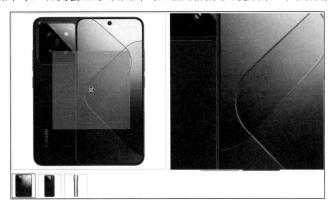

图 6.10　图片放大镜效果

具体实现步骤如下。

（1）在 views/shopinfo 文件夹下新建 ShopinfoEnlarge.vue 文件。在<template>标签中，分别定义商品图片、图片放大工具、放大的图片和商品缩略图。通过在商品图片上触发 mouseenter 事件、mouseleave 事件和 mousemove 事件，相应的方法会被执行。关键代码如下：

```
<template>
  <div class="clearfixLeft" id="clearcontent">
    <div class="box">
      <div class="enlarge" @mouseenter="mouseEnter" @mouseleave="mouseLeave" @mousemove="mouseMove">
        <img :src="bigImgUrl[n]" title="细节展示放大镜特效">
        <span class="tool"></span>
        <div class="bigbox">
          <img :src="bigImgUrl[n]" class="bigimg">
        </div>
      </div>
      <ul class="tb-thumb" id="thumblist">
        <li :class="{selected:n == index}" v-for="(item,index) in smallImgUrl" :key="index" @mouseover="setIndex(index)">
          <div class="tb-pic tb-s40">
            <a href="javascript:void(0)"><img :src="item"></a>
          </div>
        </li>
      </ul>
    </div>
    <div class="clear"></div>
  </div>
</template>
```

（2）在<script>标签中，编写鼠标在商品图片上移入、移出和移动时执行的方法。在 mouseEnter()方法中，设置图片放大工具和放大的图片显示；在 mouseLeave()方法中，设置图片放大工具和放大的图片隐藏；在 mouseMove()方法中，通过元素的定位属性设置图片放大工具和放大的图片的位置，实现图片的放大效果。关键代码如下：

```
<script>
  export default {
    data: function(){
```

```
        return {
            n: 0,                                           //缩略图索引
            smallImgUrl: [                                  //缩略图数组
                require('@/assets/images/01_small.jpg'),
                require('@/assets/images/02_small.jpg'),
                require('@/assets/images/03_small.jpg')
            ],
            bigImgUrl: [                                    //商品图片数组
                require('@/assets/images/01.jpg'),
                require('@/assets/images/02.jpg'),
                require('@/assets/images/03.jpg')
            ]
        }
    },
    methods: {
        mouseEnter: function () {                           //鼠标进入图片的效果
            document.querySelector('.tool').style.display='block';
            document.querySelector('.bigbox').style.display='block';
        },
        mouseLeave: function () {                           //鼠标移出图片的效果
            document.querySelector('.tool').style.display='none';
            document.querySelector('.bigbox').style.display='none';
        },
        mouseMove: function (e) {
            var enlarge=document.querySelector('.enlarge');
            var tool=document.querySelector('.tool');
            var bigimg=document.querySelector('.bigimg');
            var ev=window.event || e;                       //获取事件对象
            //获取图片放大工具到商品图片左端距离
            var x=ev.clientX-enlarge.offsetLeft-tool.offsetWidth/2+document.documentElement.scrollLeft;
            //获取图片放大工具到商品图片顶端距离
            var y=ev.clientY-enlarge.offsetTop-tool.offsetHeight/2+document.documentElement.scrollTop;
            if(x<0) x=0;
            if(y<0) y=0;
            if(x>enlarge.offsetWidth-tool.offsetWidth){
                x=enlarge.offsetWidth-tool.offsetWidth;     //图片放大工具到商品图片左端最大距离
            }
            if(y>enlarge.offsetHeight-tool.offsetHeight){
                y=enlarge.offsetHeight-tool.offsetHeight;   //图片放大工具到商品图片顶端最大距离
            }
            //设置图片放大工具定位
            tool.style.left = x+'px';
            tool.style.top = y+'px';
            //设置放大图片定位
            bigimg.style.left = -x * 2+'px';
            bigimg.style.top = -y * 2+'px';
        },
        setIndex: function (index) {
            this.n=index;                                   //设置缩略图索引
        }
    }
}
</script>
```

6.5.3 商品概要功能的实现

商品概要功能包含商品的名称、价格和配送地址等信息。用户通过快速浏览商品概要信息，能够迅速了解商品的销量、可配送地址和库存等内容，这有助于用户快速做出决策，节省浏览商品的时间。商品概要信息的展示效果如图 6.11 所示。

图 6.11 商品概要信息

具体实现步骤如下。

（1）在 views/shopinfo 文件夹下新建 ShopinfoInfo.vue 文件。在<template>标签中，使用<h1>标签显示商品名称，使用<li>标签显示价格信息。然后，利用<select>和<option>标签显示配送地址信息。关键代码如下：

```
<template>
  <div>
    <ol class="mr-breadcrumb mr-breadcrumb-slash">
      <li><a href="javascript:void(0)">首页</a></li>
      <li><a href="javascript:void(0)">分类</a></li>
      <li class="mr-active">内容</li>
    </ol>
    <div class="scoll">
      <section class="slider">
        <div class="flexslider">
          <ul class="slides">
            <li>
              <img src="@/assets/images/01.jpg" title="pic">
            </li>
            <li>
              <img src="@/assets/images/02.jpg">
            </li>
            <li>
              <img src="@/assets/images/03.jpg">
            </li>
          </ul>
        </div>
      </section>
    </div>
    <!--放大镜-->
    <div class="item-inform">
      <ShopinfoEnlarge/>
      <div class="clearfixRight">
        <!--规格属性-->
        <!--名称-->
        <div class="tb-detail-hd">
          <h1>
```

```html
            {{goodsInfo.name}}
          </h1>
        </div>
        <div class="tb-detail-list">
          <!--价格-->
          <div class="tb-detail-price">
            <li class="price iteminfo_price">
              <dt>促销价</dt>
              <dd><em>¥</em><b class="sys_item_price">{{goodsInfo.unitPrice.toFixed(2)}}</b></dd>
            </li>
            <li class="price iteminfo_mktprice">
              <dt>原价</dt>
              <dd><em>¥</em><b class="sys_item_mktprice">4599.00</b></dd>
            </li>
            <div class="clear"></div>
          </div>
          <!-- 省略部分代码 -->
</template>
```

（2）在<script>标签中引入 mapState()和 mapActions()辅助函数，以实现组件中的计算属性、方法和 store 中的 state、action 之间的映射。这些映射会判断用户是否登录，并根据用户的登录状态决定跳转到相应的页面。关键代码如下：

```javascript
<script>
    import ShopinfoEnlarge from '@/views/shopinfo/ShopinfoEnlarge'
    import {mapState,mapActions} from 'vuex'              //引入 mapState 和 mapActions
    export default {
        components: {
            ShopinfoEnlarge
        },
        data: function(){
            return {
                number: 1,                                //商品数量
                goodsInfo: {                              //商品基本信息
                    img : require("@/assets/images/01.jpg"),
                    name : "小米 14 徕卡光学镜头 光影猎人 900 5G 手机",
                    num : 0,
                    unitPrice : 4299,
                    isSelect : true
                }
            }
        },
        computed: {
            ...mapState([
                'user'                                    //将 this.user 映射为 this.$store.state.user
            ])
        },
        watch: {
            number: function (newVal,oldVal) {
                if(isNaN(newVal) || newVal == 0){         //输入的是非数字或 0
                    this.number = oldVal;                 //数量为原来的值
                }
            }
        },
        methods: {
            ...mapActions([
                'getListAction'                           //将 this.getListAction()映射为 this.$store.dispatch('getListAction')
            ]),
            show: function () {
                if(this.user == null){
                    alert('亲，请登录！');
                    this.$router.push({name:'login'});    //跳转到登录页面
```

```
                }else{
                    this.getListAction({                    //执行方法并传递参数
                        goodsInfo: this.goodsInfo,
                        number: parseInt(this.number)
                    });
                    this.$router.push({name:'shopcart'});   //跳转到购物车页面
                }
            },
            reduce: function () {
                if(this.number >= 2){
                    this.number--;                          //商品数量减1
                }
            },
            add: function () {
                this.number++;                              //商品数量加1
            }
        }
    }
</script>
```

6.5.4 猜你喜欢功能的实现

猜你喜欢功能旨在为用户推荐最佳相似商品，这不仅使用户能够挑选商品，还丰富了商品详情页面的内容，从而提升了用户体验。"猜你喜欢"的页面效果如图6.12所示。

图6.12 "猜你喜欢"的页面效果

具体实现步骤如下。

（1）在 views/shopinfo 文件夹下新建 ShopinfoLike.vue 文件。在<template>标签中编写商品列表区域的HTML 布局代码。首先使用<li>标签显示商品基本信息，包括商品缩略图、商品价格和商品名称等内容，然后使用<li>标签对商品信息进行分页处理。关键代码如下：

```
<template>
  <div id="youLike" class="mr-tab-panel">
    <div class="like">
      <ul class="mr-avg-sm-2 mr-avg-md-3 mr-avg-lg-4 boxes">
        <li>
          <div class="i-pic limit">
            <img src="@/assets/images/phone3.jpg">
            <p>华为 Mate60 雅丹黑 12GB+512GB 全网通手机</p>
            <p class="price fl">
              <b>¥</b>
```

```html
            <strong>7299.00</strong>
          </p>
        </div>
      </li>
      <!-- 省略部分代码 -->
    </ul>
  </div>
  <div class="clear"></div>
  <!--分页 -->
  <ul class="mr-pagination mr-pagination-right">
    <li :class="{'mr-disabled':curentPage==1}" @click="jump(curentPage-1)"><a href="javascript:void(0)">&laquo;</a></li>
    <li :class="{'mr-active':curentPage==n}" v-for="n in pages" :key="n" @click="jump(n)">
      <a href="javascript:void(0)">{{n}}</a>
    </li>
    <li :class="{'mr-disabled':curentPage==pages}" @click="jump(curentPage+1)">
      <a href="javascript:void(0)">&raquo;</a>
    </li>
  </ul>
  <div class="clear"></div>
</div>
</template>
```

（2）在<script>标签中编写实现商品信息分页的逻辑代码。在 data 选项中定义每页显示的元素个数，并通过计算属性获取元素总数和总页数。在 methods 选项中定义 jump()方法，该方法通过控制页面元素的隐藏和显示实现商品信息分页的效果。关键代码如下：

```javascript
<script>
export default {
  data: function () {
    return {
      items: [],
      eachNum: 4,                                           //每页显示个数
      curentPage: 1                                         //当前页数
    }
  },
  mounted: function(){
    this.items = document.querySelectorAll('.like li');     //获取所有元素
    for(var i = 0; i < this.items.length; i++){
      if(i < this.eachNum){
        this.items[i].style.display = 'block';              //显示第一页内容
      }else{
        this.items[i].style.display = 'none';               //隐藏其他页内容
      }
    }
  },
  computed: {
    count: function () {
      return this.items.length;                             //元素总数
    },
    pages: function () {
      return Math.ceil(this.count/this.eachNum);            //总页数
    }
  },
  methods: {
    jump: function (n) {
      this.curentPage = n;
      if(this.curentPage < 1){
        this.curentPage = 1;                                //页数最小值
      }
      if(this.curentPage > this.pages){
        this.curentPage = this.pages;                       //页数最大值
      }
```

```
            for(var i = 0; i < this.items.length; i++){
                this.items[i].style.display = 'none';           //隐藏所有元素
            }
            var start = (this.curentPage - 1) * this.eachNum;   //每页第一个元素索引
            var end = start + this.eachNum;                     //每页最后一个元素索引
            end = end > this.count ? this.count : end;          //尾页最后一个元素索引
            for(var j = start; j < end; j++){
                this.items[j].style.display = 'block';          //显示当前页元素
            }
        }
    }
}
</script>
```

6.5.5 选项卡切换效果的实现

在商品详情页面有"宝贝详情""全部评价"和"猜你喜欢"3个选项卡。当用户单击其中任意一个选项卡时,页面下方内容会立即切换为该选项卡所对应的内容,效果如图6.13所示。

图 6.13 选项卡的切换

具体实现步骤如下。

(1)在 views/shopinfo 文件夹下新建 ShopinfoIntroduce.vue 文件。在<template>标签中,首先定义"宝贝详情""全部评价"和"猜你喜欢"这3个选项卡,然后通过使用动态组件<component>,将 data 中的 current 数据动态绑定到<component>元素的 is 属性上。这允许根据 current 值的不同来显示不同的组件内容。代码如下:

```
<template>
  <div class="introduceMain">
    <div class="mr-tabs" data-mr-tabs>
      <ul class="mr-avg-sm-3 mr-tabs-nav mr-nav mr-nav-tabs">
        <li id="infoTitle" :class="{'mr-active':current=='ShopinfoDetails'}">
          <a @click="current='ShopinfoDetails'">
            <span class="index-needs-dt-txt">宝贝详情</span></a>
        </li>
        <li id="commentTitle" :class="{'mr-active':current=='ShopinfoComment'}">
          <a @click="current='ShopinfoComment'">
```

```
            <span class="index-needs-dt-txt">全部评价</span></a>
        </li>
        <li id="youLikeTitle" :class="{'mr-active':current=='ShopinfoLike'}">
          <a @click="current='ShopinfoLike'">
            <span class="index-needs-dt-txt">猜你喜欢</span></a>
        </li>
      </ul>
      <div class="mr-tabs-bd">
        <component :is="current"></component>
      </div>
    </div>
    <div class="clear"></div>
    <div class="footer ">
      <div class="footer-hd ">
        <p>
          <a href="http://www.mingrisoft.com/" target="_blank">明日科技</a>
          <b>|</b>
          <a href="javascript:void(0)" @click="show">商城首页</a>
          <b>|</b>
          <a href="javascript:void(0)">支付宝</a>
          <b>|</b>
          <a href="javascript:void(0)">物流</a>
        </p>
      </div>
      <div class="footer-bd ">
        <p>
          <a href="http://www.mingrisoft.com/Index/ServiceCenter/aboutus.html" target="_blank">关于明日</a>
          <a href="javascript:void(0)">合作伙伴</a>
          <a href="javascript:void(0)">联系我们</a>
          <a href="javascript:void(0)">网站地图</a>
          <em>&copy; 2016-2025 mingrisoft.com  版权所有</em> </p>
      </div>
    </div>
  </div>
</template>
```

（2）在<script>标签中引入上述 3 个选项卡内容对应的组件，并使用 components 选项注册这 3 个组件。关键代码如下：

```
<script>
  import ShopinfoDetails from '@/views/shopinfo/ShopinfoDetails'    //引入组件
  import ShopinfoComment from '@/views/shopinfo/ShopinfoComment'    //引入组件
  import ShopinfoLike from '@/views/shopinfo/ShopinfoLike'          //引入组件
  export default {
    name: 'ShopinfoIntroduce',
    data: function(){
      return {
        current: 'ShopinfoDetails'                                  //当前显示的组件
      }
    },
    components: {
      ShopinfoDetails,
      ShopinfoComment,
      ShopinfoLike
    },
    methods: {
      show: function () {
        this.$router.push({name:'home'});                           //跳转到主页
      }
    }
  }
</script>
```

6.6 购物车页面的设计与实现

6.6.1 购物车页面的设计

电商网站都具有购物车的功能，用户一般先将自己挑选好的商品放到购物车中，然后统一进行付款。在淘贝电子商城中，用户必须先登录账户，才可以进入购物车页面。购物车页面要求包含订单商品的型号、数量和价格等信息，以方便用户可以统一确认购买。购物车页面的效果如图 6.14 所示。

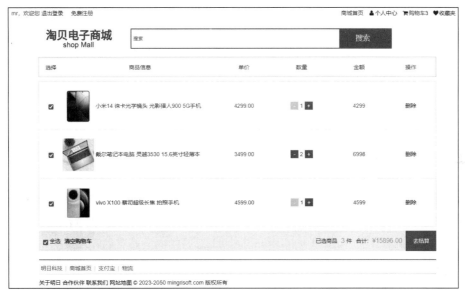

图 6.14 购物车页面的效果

6.6.2 购物车页面的实现

购物车页面分为顶部、主显示区和底部 3 个部分。这里重点讲解购物车页面中主显示区的实现方法。具体实现步骤如下。

（1）在 views/shopcart 文件夹下新建 ShopcartCart.vue 文件。在<template>标签中使用 v-for 指令循环输出购物车中的商品信息。在商品数量一栏中添加"-"按钮和"+"按钮，用户单击这两个按钮时执行相应的方法实现商品数量减 1 或加 1 的操作。在操作一栏中添加"删除"超链接，用户单击该超链接时会执行 remove() 方法，实现删除指定商品的操作。关键代码如下：

```
<template>
  <div>
    <div v-if="list.length>0">
      <div class="main">
        <div class="goods" v-for="(item,index) in list" :key="index">
          <span class="check"><input type="checkbox" @click="selectGoods(index)" :checked="item.isSelect"> </span>
          <span class="name"><img :src="item.img">{{item.name}}</span>
          <span class="unitPrice">{{item.unitPrice.toFixed(2)}}</span>
          <span class="num">
            <span @click="reduce(index)" :class="{off:item.num==1}">-</span>
            {{item.num}}
```

```
            <span @click="add(index)">+</span>
          </span>
          <span class="unitTotalPrice">{{item.unitPrice * item.num.toFixed(2)}}</span>
          <span class="operation">
            <a @click="remove(index)">删除</a>
          </span>
        </div>
      </div>
      <div class="info">
        <span><input type="checkbox" @click="selectAll" :checked="isSelectAll"> 全选</span>
        <a @click="emptyCar">清空购物车</a>
        <span>已选商品<span class="totalNum">{{totalNum}}</span> 件</span>
        <span>合计:<span class="totalPrice">¥{{totalPrice.toFixed(2)}}</span></span>
        <span @click="show('pay')">去结算</span>
      </div>
    </div>
    <div class="empty" v-else>
      <img src="@/assets/images/shopcar.png">
      购物车内暂时没有商品，<a @click="show('home')">去购物></a>
    </div>
  </div>
</template>
```

（2）在<script>标签中引入 mapState()和 mapActions()辅助函数，以实现组件中的计算属性、方法和 store 中的 state、action 之间的映射。计算属性将用于统计选择的商品件数和商品总价，在 methods 选项中，定义不同的方法实现选择某个商品、全选商品和跳转到指定页面等操作。关键代码如下：

```
<script>
  import { mapState,mapActions } from 'vuex'              //引入 mapState 和 mapActions
  export default{
    data: function () {
      return {
        isSelectAll : false                                //默认未全选
      }
    },
    mounted: function(){
      this.isSelectAll = true;                             //全选
      for(var i = 0;i < this.list.length; i++){
        //有一个商品未选中即取消全选
        if(this.list[i].isSelect == false){
          this.isSelectAll=false;
        }
      }
    },
    computed : {
      ...mapState([
        'list'                                             //将 this.list 映射为 this.$store.state.list
      ]),
      totalNum : function(){                               //计算商品件数
        var totalNum = 0;
        this.list.forEach(function(item){
          if(item.isSelect){
            totalNum+=1;
          }
        });
        return totalNum;
      },
      totalPrice : function(){                             //计算商品总价
        var totalPrice = 0;
        this.list.forEach(function(item){
          if(item.isSelect){
            totalPrice += item.num*item.unitPrice;
```

```
          }
       });
       return totalPrice;
     }
  },
  methods : {
    ...mapActions({
      reduce: 'reduceAction',              //减少商品个数
      add: 'addAction',                    //增加商品个数
      remove: 'removeGoodsAction',         //移除商品
      selectGoodsAction: 'selectGoodsAction',  //选择商品
      selectAllAction: 'selectAllAction',  //全选商品
      emptyCarAction: 'emptyCarAction'     //清空购物车
    }),
    selectGoods : function(index){         //选择商品
      var goods = this.list[index];
      goods.isSelect = !goods.isSelect;
      this.isSelectAll = true;
      for(var i = 0;i < this.list.length; i++){
        if(this.list[i].isSelect == false){
          this.isSelectAll=false;
        }
      }
      this.selectGoodsAction({
        index: index,
        bool: goods.isSelect
      });
    },
    selectAll : function(){                //全选或全不选
      this.isSelectAll = !this.isSelectAll;
      this.selectAllAction(this.isSelectAll);
    },
    emptyCar: function(){                  //清空购物车
      if(confirm('确定要清空购物车吗？')){
        this.emptyCarAction();
      }
    },
    show: function (value) {
      if(value == 'home'){
        this.$router.push({name:'home'});  //跳转到主页
      }else{
        if(this.totalNum==0){
          alert('请至少选择一件商品！');
          return false;
        }
        this.$router.push({name:'pay'});   //跳转到支付页面
      }
    }
  }
}
</script>
```

6.7 付款页面的设计与实现

6.7.1 付款页面的设计

用户在购物车页面单击"去结算"按钮后，进入付款页面。付款页面包括收货人姓名、手机号、收货

地址、物流方式和支付方式等内容。用户需要再次确认上述内容后，单击"提交订单"按钮，完成交易。付款页面的效果如图 6.15 所示。

图 6.15　付款页面效果

6.7.2　付款页面的实现

付款页面包括多个组件。这里重点讲解付款页面中确认订单信息的组件 PayOrder.vue 和执行订单提交的组件 PayInfo.vue。确认订单信息页面的效果如图 6.16 所示。执行订单提交页面的效果如图 6.17 所示。

图 6.16　确认订单信息页面效果　　　　图 6.17　执行订单提交页面效果

PayOrder.vue 组件的具体实现步骤如下。

（1）在 views/pay 文件夹下新建 PayOrder.vue 文件。在<template>标签中使用 v-for 指令循环输出购物车中选中的商品信息，包括商品名称、单价、数量、金额等。关键代码如下：

```
<template>
  <!--订单 -->
  <div>
    <div class="concent">
      <div id="payTable">
        <h3>确认订单信息</h3>
        <div class="cart-table-th">
          <div class="wp">
            <div class="th th-item">
              <div class="td-inner">商品信息</div>
            </div>
            <div class="th th-price">
              <div class="td-inner">单价</div>
            </div>
            <div class="th th-amount">
              <div class="td-inner">数量</div>
            </div>
            <div class="th th-sum">
              <div class="td-inner">金额</div>
            </div>
            <div class="th th-oplist">
              <div class="td-inner">配送方式</div>
            </div>
          </div>
        </div>
        <div class="clear"></div>
        <div class="main">
          <div class="goods" v-for="(item,index) in list" :key="index">
            <span class="name">
              <img :src="item.img">
              {{item.name}}
            </span>
            <span class="unitPrice">{{item.unitPrice.toFixed(2)}}</span>
            <span class="num">
              {{item.num}}
            </span>
            <span class="unitTotalPrice">{{item.unitPrice * item.num.toFixed(2)}}</span>
            <span class="pay-logis">
              快递送货
            </span>
          </div>
```

```
        </div>
      </div>
    </div>
    <PayMessage :totalPrice="totalPrice"/>
  </div>
</template>
```

（2）在<script>标签中引入 mapGetters()辅助函数，实现组件中的计算属性和 store 中的 getter 之间的映射。通过计算属性获取购物车中选中的商品，以及统计单个商品的总价。关键代码如下：

```
<script>
  import {mapGetters} from 'vuex'                              //引入 mapGetters
  import PayMessage from '@/views/pay/PayMessage'              //引入组件
  export default {
    components:{
      PayMessage                                               //注册组件
    },
    computed: {
      ...mapGetters([
        'list'                                                 //this.list 映射为 this.$store.getters.list
      ]),
      totalPrice : function(){                                 //计算商品总价
        var totalPrice = 0;
        this.list.forEach(function(item){
          if(item.isSelect){
            totalPrice += item.num*item.unitPrice;
          }
        });
        return totalPrice;
      }
    }
  }
</script>
```

PayInfo.vue 组件的具体实现步骤如下。

（1）在 views/pay 文件夹下新建 PayInfo.vue 文件。在<template>标签中定义实付款、收货地址以及收货人信息，并设置当单击"提交订单"按钮时执行 show()方法。关键代码如下：

```
<template>
  <!--信息 -->
  <div class="order-go clearfix">
    <div class="pay-confirm clearfix">
      <div class="box">
        <div tabindex="0" id="holyshit267" class="realPay"><em class="t">实付款： </em>
          <span class="price g_price ">
            <span>¥</span>
            <em class="style-large-bold-red " id="J_ActualFee">{{lastPrice.toFixed(2)}}</em>
          </span>
        </div>
        <div id="holyshit268" class="pay-address">
          <p class="buy-footer-address">
            <span class="buy-line-title buy-line-title-type">寄送至： </span>
            <span class="buy--address-detail">
              <span class="province">吉林</span>省
              <span class="city">长春</span>市
              <span class="dist">朝阳</span>区
              <span class="street">**花园****号</span>
            </span>
          </p>
          <p class="buy-footer-address">
            <span class="buy-line-title">收货人： </span>
            <span class="buy-address-detail">
              <span class="buy-user">Tony </span>
```

```
            <span class="buy-phone">1567699****</span>
          </span>
        </p>
      </div>
    </div>
    <div id="holyshit269" class="submitOrder">
      <div class="go-btn-wrap">
        <a id="J_Go" class="btn-go" tabindex="0" title="点击此按钮，提交订单" @click="show">提交订单</a>
      </div>
    </div>
    <div class="clear"></div>
  </div>
</div>
</template>
```

（2）在<script>标签中引入 mapActions()辅助函数，实现组件中的方法和 store 中的 action 之间的映射。在 methods 选项中定义 show()方法，在方法中执行清空购物车的操作，并通过路由跳转到商城主页。关键代码如下：

```
<script>
  import {mapActions} from 'vuex'              //引入 mapActions
  export default {
    props:['lastPrice'],                       //父组件传递的数据
    methods: {
      ...mapActions({
        emptyCar: 'emptyCarAction'             //将 this.emptyCar()映射为 this.$store.dispatch('emptyCarAction')
      }),
      show: function () {
        this.emptyCar();                       //执行清空购物车操作
        this.$router.push({name:'home'});      //跳转到主页
      }
    }
  }
</script>
```

6.8　注册和登录页面的设计与实现

6.8.1　注册和登录页面的设计

注册和登录页面是通用的功能页面。淘贝电子商城在设计注册和登录页面时，使用简单的 JavaScript 方法验证邮箱和数字的格式。注册和登录页面的效果分别如图 6.18 和图 6.19 所示。

图 6.18　注册页面效果

6.8.2 注册页面的实现

注册页面分为顶部、主显示区和底部 3 个部分。这里重点讲解主显示区中注册页面的布局以及用户注册信息的验证。在验证用户输入的表单信息时，需要验证邮箱格式是否正确、手机格式是否正确等。注册页面的效果如图 6.20 所示。

图 6.19　登录页面效果　　　　　　　　图 6.20　注册页面效果

具体实现步骤如下。

（1）在 views/register 文件夹下新建 RegisterHome.vue 文件。在<template>标签中编写注册页面的 HTML 代码。首先定义用户注册的表单信息，并使用 v-model 指令对表单元素进行数据绑定，然后通过<input>标签设置一个"注册"按钮，当单击该按钮时会执行 mr_verify()方法。关键代码如下：

```
<template>
  <div>
    <div class="res-banner">
      <div class="res-main">
        <div class="login-banner-bg"><span></span><img src="@/assets/images/big.jpg"/></div>
        <div class="login-box">
          <div class="mr-tabs" id="doc-my-tabs">
            <h3 class="title">注 册</h3>
            <div class="mr-tabs-bd">
              <div class="mr-tab-panel mr-active">
                <form method="post">
                  <div class="user-email">
                    <label for="email"><i class="mr-icon-envelope-o"></i></label>
                    <input type="email" v-model="email" id="email" placeholder="请输入邮箱账号">
                  </div>
                  <div class="user-pass">
                    <label for="password"><i class="mr-icon-lock"></i></label>
                    <input type="password" v-model="password" id="password" placeholder="设置密码">
                  </div>
                  <div class="user-pass">
                    <label for="passwordRepeat"><i class="mr-icon-lock"></i></label>
                    <input type="password" v-model="passwordRepeat" id="passwordRepeat" placeholder="确认密码">
                  </div>
                  <div class="user-pass">
                    <label for="passwordRepeat">
                      <i class="mr-icon-mobile"></i>
                      <span style="color:red;margin-left:5px">*</span>
                    </label>
                    <input type="text" v-model="tel" id="tel" placeholder="请输入手机号">
                  </div>
```

```html
      </form>
      <div class="login-links">
        <label for="reader-me">
          <input id="reader-me" type="checkbox" v-model="checked">  选中以表示您同意商城《服务协议》
        </label>
        <a href="javascript:void(0)" @click="show" class="mr-fr">登录</a>
      </div>
      <div class="mr-cf">
        <input type="submit" name="" :disabled="!checked" @click="mr_verify" value=
            "注册" class="mr-btn mr-btn-primary mr-btn-sm mr-fl">
      </div>
     </div>
    </div>
   </div>
  </div>
  <RegisterBottom/>
 </div>
</template>
```

（2）在<script>标签中编写验证用户注册信息的代码。在 data 选项中定义注册表单元素绑定的数据，然后在 methods 选项中定义 mr_verify()方法，在该方法中分别获取用户输入的邮箱、密码、确认密码和手机号码信息，并验证用户输入是否正确。如果输入正确，则弹出相应的提示信息，并跳转到商城主页。代码如下：

```javascript
<script>
  import RegisterBottom from '@/views/register/RegisterBottom'    //引入组件
  export default {
    name : 'RegisterHome',
    components : {
      RegisterBottom                                              //注册组件
    },
    data: function(){
      return {
        checked:false,                                            //是否同意注册协议复选框
        email:'',                                                 //电子邮箱
        password:'',                                              //密码
        passwordRepeat:'',                                        //确认密码
        tel:''                                                    //手机号
      }
    },
    methods: {
      mr_verify: function () {
        //获取表单对象
        var email=this.email;
        var password=this.password;
        var passwordRepeat=this.passwordRepeat;
        var tel=this.tel;
        //验证表单元素是否为空
        if(email==='' || email===null){
          alert("邮箱不能为空！");
          return;
        }
        if(password==='' || password===null){
          alert("密码不能为空！");
          return;
        }
        if(passwordRepeat==='' || passwordRepeat===null){
          alert("确认密码不能为空！");
          return;
        }
        if(tel==='' || tel===null){
```

```
            alert("手机号码不能为空！");
            return;
          }
          if(password!==passwordRepeat){
            alert("密码设置前后不一致！");
            return;
          }
          //验证邮件格式
          var apos = email.indexOf("@")
          var dotpos = email.lastIndexOf(".")
          if (apos < 1 || dotpos - apos < 2) {
            alert("邮箱格式错误！");
            return;
          }
          //验证手机号格式
          if(!/^1[345789]\d{9}$/.test(tel)){
            alert("手机号格式错误！");
            return;
          }
          alert('注册成功！');
          this.$router.push({name:'home'});                    //跳转到主页
        },
        show: function () {
          this.$router.push({name:'login'});                    //跳转到登录页面
        }
      }
    }
</script>
```

> **说明**
> JavaScript 验证手机号格式是否正确的原理是利用正则表达式（Regular Expression）来验证数字（或数字字符）的格式是否符合手机号的标准。

6.8.3 登录页面的实现

登录页面和注册页面一样，同样分为顶部、主显示区和底部 3 个部分。这里重点讲解主显示区中登录页面的布局和用户登录的验证。登录页面的效果如图 6.21 所示。

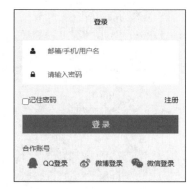

图 6.21　登录页面效果

具体实现步骤如下。

（1）在 views/login 文件夹下新建 LoginHome.vue 文件。在<template>标签中编写登录页面的 HTML 代码。首先定义用于显示用户名和密码的表单，并使用 v-model 指令对表单元素进行数据绑定，然后通过<input>标签设置一个"登录"按钮，当用户单击该按钮时会执行 login()方法。关键代码如下：

```
<template>
  <div>
    <div class="login-banner">
      <div class="login-main">
        <div class="login-banner-bg"><span></span><img src="@/assets/images/big.jpg"/></div>
        <div class="login-box">
          <h3 class="title">登录</h3>
          <div class="clear"></div>
          <div class="login-form">
            <form>
```

```html
            <div class="user-name">
              <label for="user"><i class="mr-icon-user"></i></label>
              <input type="text" v-model="user" id="user" placeholder="邮箱/手机/用户名">
            </div>
            <div class="user-pass">
              <label for="password"><i class="mr-icon-lock"></i></label>
              <input type="password" v-model="password" id="password" placeholder="请输入密码">
            </div>
          </form>
        </div>
        <div class="login-links">
          <label for="remember-me"><input id="remember-me" type="checkbox">记住密码</label>
          <a href="javascript:void(0)" @click="show" class="mr-fr">注册</a>
          <br/>
        </div>
        <div class="mr-cf">
          <input type="submit" name="" value="登  录" @click="login" class="mr-btn mr-btn-primary mr-btn-sm">
        </div>
        <div class="partner">
          <h3>合作账号</h3>
          <div class="mr-btn-group">
            <li><a href="javascript:void(0)"><i class="mr-icon-qq mr-icon-sm"></i><span>QQ 登录</span></a></li>
            <li><a href="javascript:void(0)"><i class="mr-icon-weibo mr-icon-sm"></i><span>微博登录</span> </a></li>
            <li><a href="javascript:void(0)"><i class="mr-icon-weixin mr-icon-sm"></i><span>微信登录</span> </a></li>
          </div>
        </div>
      </div>
    </div>
    <LoginBottom/>
  </div>
</template>
```

（2）在<script>标签中编写验证用户登录的代码。首先引入 mapActions()辅助函数，以实现组件中的方法和 store 中的 action 之间的映射。在 methods 选项中定义 login()方法，在该方法中分别获取用户输入的用户名和密码信息，并验证用户输入是否正确。如果输入正确，则弹出相应的提示信息，接着执行 loginAction()方法对用户名进行存储，并跳转到商城主页。代码如下：

```javascript
<script>
  import {mapActions} from 'vuex'                              //引入 mapActions
  import LoginBottom from '@/views/login/LoginBottom'          //引入组件
  export default {
    name : 'LoginHome',
    components : {
      LoginBottom                                              //注册组件
    },
    data: function(){
      return {
        user:null,                                             //用户名
        password:null                                          //密码
      }
    },
    methods: {
      ...mapActions([
        'loginAction'                                          //将 this.loginAction()映射为 this.$store.dispatch('loginAction')
      ]),
      login: function () {
        var user=this.user;                                    //获取用户名
        var password=this.password;                            //获取密码
        if(user == null){
```

```
            alert('请输入用户名！');
            return false;
        }
        if(password == null){
            alert('请输入密码！');
            return false;
        }
        if(user!=='mr' || password!=='mrsoft' ){
            alert('您输入的账户或密码错误！');
            return false;
        }else{
            alert('登录成功！');
            this.loginAction(user);                         //触发 action 并传递用户名
            this.$router.push({name:'home'});               //跳转到主页
        }
      },
      show: function () {
        this.$router.push({name:'register'});               //跳转到注册页面
      }
    }
  }
</script>
```

> **说明**
> 默认正确用户名为 mr，密码为 mrsoft。如果输入错误，则系统会提示"您输入的账户或密码错误"；如果输入正确，则系统会提示"登录成功"。

6.9 项目运行

通过前述步骤，我们已经设计并完成了"淘贝电子商城"项目的开发。接下来，我们运行该项目，以检验我们的开发成果。首先打开命令提示符窗口，切换到项目所在目录，执行 npm run serve 命令运行该项目，如图 6.22 所示。

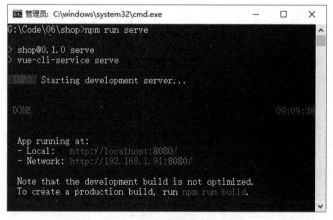

图 6.22 运行项目

在浏览器地址栏中输入 http://localhost:8080，然后按 Enter 键，即可成功运行该项目。淘贝电子商城的主页效果如图 6.23 所示。

淘贝电子商城 第 6 章

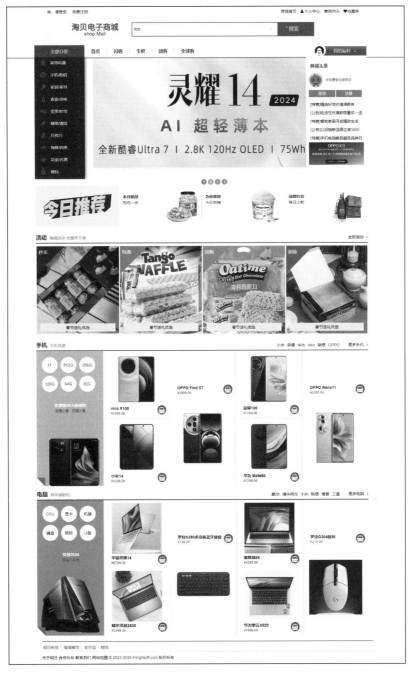

图 6.23 淘贝电子商城的主页效果

6.10 源 码 下 载

本章虽然详细地讲解了如何编码实现"淘贝电子商城"的各个功能，但给出的代码都是代码片段，而非完整的源代码。为了方便读者学习，本书提供了该项目的完整源代码，读者可以扫描右侧的二维码进行下载。

源码下载

第 7 章 畅联通讯录

——Vue CLI + Vue Router + Vuex + localStorage + sessionStorage

通讯录是一种用于存储和管理联系人信息的工具。随着互联网的普及和通信技术的发展，人们对于通讯录系统的需求越来越大。目前，通讯录系统已经广泛应用于各个行业。本章将使用 Vue CLI、Vue Router、Vuex、localStorage 和 sessionStorage 等技术开发一个通讯录管理系统——畅联通讯录。

项目微视频

本项目的核心功能及实现技术如下：

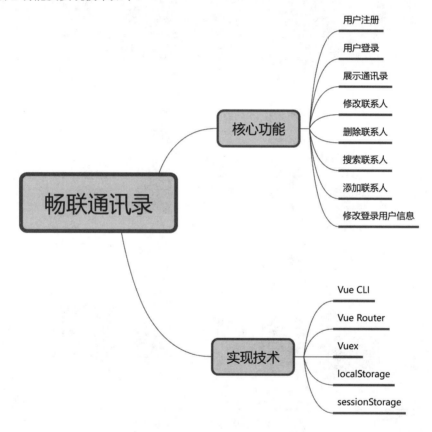

7.1 开发背景

随着互联网技术的发展，计算机已经成为人们生活中不可或缺的一部分，它打破了地域和时间的限制，改变了人们的工作和生活方式。在当今社会，人们之间的联系变得更加便捷，社交网络也变得更加庞大。单纯依靠人脑已经很难记住所有人的联系方式和其他的通讯信息，同时传统的纸质通讯录已经不能满足人们的需求。因此，人们越来越多地依赖于手机或其他电子设备进行联系，通讯录系统已经成为人们日常生活的必备软件工具。通讯录系统可以帮助用户更方便地存储和管理联系人信息，从而提升沟通效率。

本章将使用 Vue CLI、Vue Router、Vuex、localStorage 和 sessionStorage 等技术开发一个通讯录系统——畅联通讯录。项目的实现目标如下：

- ☑ 实现用户注册和登录。
- ☑ 展示通讯录列表。
- ☑ 提供联系人信息搜索功能。
- ☑ 提供联系人信息分类功能。
- ☑ 提供修改和删除联系人的功能。
- ☑ 可以向联系人列表中添加联系人。
- ☑ 可以修改登录用户的信息。

7.2 系统设计

7.2.1 开发环境

本项目的开发及运行环境如下：

- ☑ 操作系统：推荐 Windows 10、Windows 11 或更高版本，同时兼容 Windows 7（SP1）。
- ☑ 开发工具：WebStorm。
- ☑ 开发框架：Vue.js 3.0。

7.2.2 业务流程

畅联通讯录由多个页面组成，包括注册页面、登录页面、通讯录页面、添加联系人页面和个人中心页面等。根据该项目的业务需求，我们设计如图 7.1 所示的业务流程图。

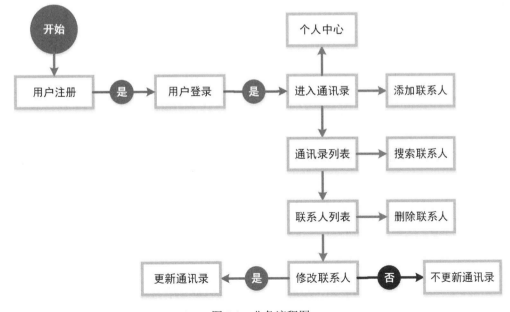

图 7.1　业务流程图

7.2.3 功能结构

本项目的功能结构已经在章首页中给出，其实现的具体功能如下：
- 用户注册：在注册时需要输入用户名、电话号码和密码，并验证用户输入的内容是否符合要求。
- 用户登录：在登录时需要输入用户名和密码，在验证时分别对用户名和密码进行判断。
- 更新通讯录：展示通讯录列表，包括联系人头像、联系人姓名和联系人电话号码。
- 修改联系人：单击联系人右侧的修改图标进行修改。
- 删除联系人：单击联系人右侧的删除图标以删除联系人。
- 搜索联系人：在通讯录列表上方的文本框中输入搜索关键字，下方将展示联系人姓名中包含该关键字的联系人列表。
- 添加联系人：单击通讯录页面左侧的"添加联系人"超链接，进入添加联系人页面，然后在此页面的表单中输入联系人姓名、电话号码，并选择该联系人和当前登录用户的关系，最后单击"添加"按钮进行添加。
- 修改个人信息：单击通讯录页面左侧的"个人中心"超链接，进入个人中心页面。在该页面中，用户可以对当前登录用户的个人信息进行修改。

7.3 技术准备

在开发畅联通讯录时，我们主要应用 Vue CLI 脚手架工具、Vue Router、Vuex、localStorage 和 sessionStorage 等技术。其中，Vue CLI 脚手架工具、Vue Router 和 Vuex 在《Vue.js 从入门到精通》中有详细的讲解，对这些知识不太熟悉的读者，可以参考该书对应的内容。localStorage 已经在 6.3 节做了简要介绍，读者可以参考对应内容。sessionStorage 和 localStorage 一样，也是 HTML5 中的内容，下面将对 sessionStorage 进行必要介绍，以确保读者可以顺利完成本项目。

1. 什么是 sessionStorage

sessionStorage 是 HTML5 新增的一个会话存储对象，用于临时保存同一窗口（或标签页）的数据，在关闭窗口或标签页之后将会删除这些数据。

sessionStorage 的特点如下：
- 同源策略限制：若想在不同页面之间对同一个 sessionStorage 进行操作，这些页面必须遵循同源策略，即它们位于同一协议、同一主机名和同一端口下。
- 单标签页限制：sessionStorage 操作限制在单个标签页中，在此标签页中进行同源页面访问都可以共享 sessionStorage 数据。
- 只在本地存储：sessionStorage 中的数据不会随着 HTTP 请求被发送到服务器，仅在本地生效，并且一旦关闭标签页存储的数据就会被清除。
- 存储方式：sessionStorage 的存储方式采用 key、value 的方式。其中，value 的值必须是字符串类型。
- 存储上限限制：不同的浏览器对 sessionStorage 的存储上限设置各不相同，大多数浏览器将其设置为 5 MB。

2. sessionStorage 常用属性和方法

- setItem(key, value)方法：该方法用于设置 key 的值为 value。key 的类型是字符串类型。value 可以是包括字符串、布尔值、整数或者浮点数在内的任意 JavaScript 支持的类型。但是，最终数据是以字符串类型存储的。如果指定的 key 已经存在，那么新传入的数据会覆盖原来的数据。例如，使用 sessionStorage 对象保存键为 userId、值为 1 的数据，代码如下：

sessionStorage.setItem("userId", "1");

- length 属性：该属性用于获取 sessionStorage 中保存数据项（item）的个数。例如，想要获取 sessionStorage 中存储的数据项的个数，可以使用下面的代码：

sessionStorage.length;

- key(index)方法：该方法用于返回某个索引（从 0 开始计数）对应的 key，即 sessionStorage 中的第 index 个数据项的键。例如，想要获取 sessionStorage 中索引为 2 的数据项的键，可以使用下面的代码：

sessionStorage.key(2);

- getItem(key)方法：该方法用于返回指定 key 的值，即存储的内容，类型为 String 字符串类型。如果传入的 key 不存在，那么该方法会返回 null，而不会抛出异常。例如，想要获取 sessionStorage 中 key 为 userId 的数据项的值，可以使用下面的代码：

sessionStorage.getItem("userId");

- removeItem(key)方法：该方法用于从存储列表中删除指定 key（包括对应的值）。例如，想要删除 sessionStorage 中 key 为 userId 的数据项，可以使用下面的代码：

sessionStorage.removeItem("userId");

- clear()方法：该方法用于清空 sessionStorage 中的全部数据项。例如，想要清空 sessionStorage 中的数据项，可以使用下面的代码：

sessionStorage.clear();

说明

sessionStorage 只能存储字符串类型的数据。如果要存储对象或数组，可以通过 JSON.stringify()方法将 JavaScript 对象转化成字符串再进行存储。在获取数据时，需要使用 JSON.parse()方法将获取的字符串转换回 JavaScript 对象。

7.4 创 建 项 目

在设计畅联通讯录各功能模块之前，需要使用 Vue CLI 创建一个新的项目，并将项目名称设置为 address-book。在命令提示符窗口中输入以下命令：

vue create address-book

按 Enter 键，选择 Manually select features，如图 7.2 所示。
按 Enter 键后，选择 Router 和 Vuex 选项，如图 7.3 所示。
按 Enter 键后，选择 Vue 3.x 版本，如图 7.4 所示。
然后选择路由是否使用 history 模式，输入 y 表示使用 history 模式，如图 7.5 所示。

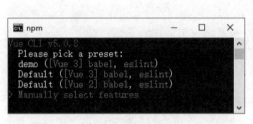

图 7.2　选择 Manually select features

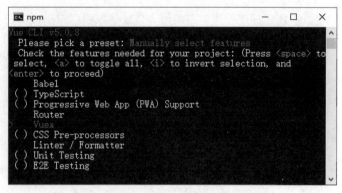

图 7.3　选择配置选项

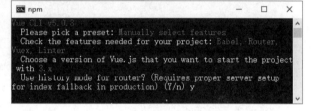

图 7.4　选择 Vue 版本　　　　　　　　　图 7.5　使用 history 模式

创建项目后，首先需要整理项目目录，然后在 assets 目录中创建一个名为 img 的文件夹，用于存储项目所需的图片文件。

7.5　注册和登录页面设计

在进入通讯录页面之前，用户需要进行登录，而在登录之前，用户如果尚未注册，则需要进行注册。注册时，用户需要输入用户名、电话号码和密码，页面效果如图 7.6 所示。登录时，用户需要输入用户名和密码，页面效果如图 7.7 所示。

图 7.6　注册页面效果　　　　　　　　　图 7.7　登录页面效果

7.5.1 页面头部组件设计

注册和登录页面的头部主要包括通讯录名称、通讯录描述，以及"注册"和"登录"超链接，其页面效果如图7.8所示。

注册和登录页面头部组件的实现过程如下。

（1）在components文件夹下创建页面头部组件HomeTop.vue。在<template>标签中，分别定义通讯录名称、通讯录描述，并使用<router-link>标签定义"注册"和"登录"超链接，同时使用<router-view>标签渲染路由组件。代码如下：

图7.8 注册和登录页面头部组件效果

```
<template>
    <div id="contain">
        <h1>畅联</h1>
        <h2>畅联通讯，连通你我</h2>
        <div class="button">
            <router-link to="/home/register" class="reg">注册</router-link>
            <router-link to="/home/login">登录</router-link>
        </div>
        <router-view></router-view>
    </div>
</template>
```

（2）在<style>标签中为页面头部组件中的元素设置样式，代码如下：

```
<style scoped>
    html {
        font-size: 10px;                                /*设置文字大小*/
    }
    li {
        list-style: none;                               /*设置列表无样式*/
    }
    html, body, #contain {
        height: 100vh;                                  /*设置高度*/
        margin: 0;                                      /*设置外边距*/
    }
    body {
        color: #555;                                    /*设置文字颜色*/
        background-color: #f7fafc;                      /*设置背景颜色*/
    }
    #contain {
        width: 100%;                                    /*设置宽度*/
        text-align: center;                             /*设置文本水平居中对齐*/
        padding-top: 92px;                              /*设置上内边距*/
        box-sizing: border-box;                         /*宽度和高度包括内容、内边距和边框*/
        background-size: cover;                         /*设置背景图像尺寸*/
    }
    #contain h1 {
        color: #0f88eb;                                 /*设置文字颜色*/
        font-weight: 500;                               /*设置文字粗细*/
        text-align: center;                             /*设置文本水平居中对齐*/
        font-size: 70px;                                /*设置文字大小*/
    }
    #contain h2 {
        margin-top: 30px;                               /*设置上外边距*/
        margin-bottom: 20px;                            /*设置下外边距*/
        font-weight: 400;                               /*设置文字粗细*/
        font-size: 18px;                                /*设置文字大小*/
```

```
            line-height: 1;                        /*设置行高*/
            text-align: center;                    /*设置文本水平居中对齐*/
        }
        #contain .button {
            font-size: 18px;                       /*设置文字大小*/
            text-align: center;                    /*设置文本水平居中对齐*/
            margin-bottom: 25px;                   /*设置下外边距*/
            box-sizing: border-box;                /*宽度和高度包括内容、内边距和边框*/
        }
        a {
            color: #555;                           /*设置文字颜色*/
            text-decoration: none;                 /*设置无下画线*/
            opacity: .7;                           /*设置不透明度*/
            font-weight: 500;                      /*设置文字粗细*/
        }
        #contain .reg {
            margin-right: 20px;                    /*设置右外边距*/
        }
        .router-link-active {
            color: #0f88eb;                        /*设置文字颜色*/
            opacity: 1;                            /*设置完全不透明*/
            border-bottom: 2px solid;              /*设置下边框*/
            padding-bottom: 5px;                   /*设置下内边距*/
        }
</style>
```

7.5.2 用户注册组件设计

在注册过程中系统会对用户输入的内容进行验证。一旦验证通过，用户单击"注册畅联"按钮，系统就会提示注册成功。注册成功的页面效果如图 7.9 所示。

用户注册组件的实现过程如下。

（1）在 components 文件夹下创建用户注册组件 UserRegister.vue。在<template>标签中，定义用户注册表单，并将表单元素和定义的数据进行绑定。当输入框获得焦点或其值发生改变时，将分别调用 judge()方法进行验证；当用户单击按钮时，将调用 toReg()方法以处理注册逻辑，代码如下：

图 7.9 注册成功的页面效果

```
<template>
    <div class="register">
        <form>
            <label for="name">
                <input type="text" class="input" id="name" @focus="judge(1)" @input="judge(1)"
                    v-model="user.name" placeholder="用户名">
                <em v-show="isShowName">*用户名只能由字母、数字或下画线组成</em>
                <em v-show="isSuc===false">*该用户名已被使用</em>
            </label>
            <label for="tel">
                <input type="text" class="input" id="tel" @focus="judge(2)" @input="judge(2)"
                    v-model="user.tel" placeholder="电话号码">
                <em v-show="isShowTel">*电话号码只能由数字组成</em>
            </label>
            <label for="pwd">
                <input type="password" class="input" id="pwd" @focus="judge(3)" @input="judge(3)"
                    v-model="user.pwd" placeholder="密码（不少于 6 位）">
                <em v-show="isShowPwd">*密码不能小于 6 位</em>
            </label>
```

```html
            <button type="button" class="registerSub" @click="toReg">
                <span v-show="!result">注册畅联</span>
                <span v-if="isSuc" v-show="result">注册成功</span>
                <span v-else v-show="result">重新注册</span>
                <i class="fa fa-spinner fa-spin" v-show="icon"></i>
                <i class="fa fa-check" v-if="isSuc" v-show="result"></i>
                <i class="fa fa-times" v-else v-show="result"></i>
            </button>
        </form>
    </div>
</template>
```

（2）在<script>标签中引入 mapMutations()辅助函数，用户将组件中的方法与 store 中的 mutation 进行映射。然后定义数据和方法。judge()方法用于判断用户输入的内容是否符合要求。toReg()方法将进行最终的判断，如果三个输入框的输入内容都符合要求，它将调用 register()方法进行注册。代码如下：

```javascript
<script>
    import { mapMutations } from 'vuex'                          //引入 mapMutations
    export default {
        data() {
            return {
                user: {
                    name: '',
                    tel: '',
                    pwd: '',
                },
                icon: false,                                     //控制是否显示旋转图标
                result: false,                                   //控制是否显示响应图标
                isSuc: '',                                       //控制是否注册成功
                isShowName: false,                               //控制是否显示用户名提示信息
                isShowTel: false,                                //控制是否显示电话号码提示信息
                isShowPwd: false                                 //控制是否显示密码提示信息
            };
        },
        methods: {
            ...mapMutations(['register']),
            judge(n){
                if(n === 1){
                    //判断用户名是否符合要求
                    this.isShowName = this.user.name === '' || /\W/g.test(this.user.name);
                } else if (n === 2){
                    //判断电话号码是否符合要求
                    this.isShowTel = this.user.tel==='' || /\D/g.test(this.user.tel);
                } else {
                    //判断密码是否符合要求
                    this.isShowPwd = this.user.pwd.length < 6;
                }
            },
            toReg() {
                //判断三个输入框是否都有值
                if(this.user.name && this.user.tel && this.user.pwd){
                    //判断三个输入框的输入内容是否都符合要求
                    if(!this.isShowName && !this.isShowTel && !this.isShowPwd) {
                        this.icon = true;
                        this.result = false;
                        this.isSuc = '';
                        setTimeout(() => {
                            this.result = true;
                            this.icon = false;
                            this.register(this.user);
                            //判断是否注册成功
                            if(sessionStorage.getItem('register') === '1') {
```

```
                            this.isSuc = true;
                            this.user.name = "";
                            this.user.tel = '';
                            this.user.pwd = '';
                        } else
                            this.isSuc = false;
                    }, 1000);
                }
            }
        }
    }
</script>
```

（3）在<style>标签中为用户注册组件中的元素设置样式，代码如下：

```
<style scoped>
    .register {
        width: 994px;                                       /*设置宽度*/
        margin: 0 auto;                                     /*设置外边距*/
    }
    .register label {
        padding-left: 347px;                                /*设置左内边距*/
        font-size: 15px;                                    /*设置文字大小*/
        display: block;                                     /*设置元素为块级元素*/
        text-align: left;                                   /*设置文本水平左对齐*/
    }
    .register label em {
        margin-left: 5px;                                   /*设置左外边距*/
    }
    .register .input {
        width: 300px;                                       /*设置宽度*/
        height: 47px;                                       /*设置高度*/
        line-height: 47px;                                  /*设置行高*/
        margin-bottom: 20px;                                /*设置下外边距*/
        border: 1px solid #d5d5d5;                          /*设置边框*/
        box-shadow: none;                                   /*设置框无阴影*/
        padding: 0 8px;                                     /*设置内边距*/
        background-color: #fff;                             /*设置背景颜色*/
        box-sizing: border-box;                             /*宽度和高度包括内容、内边距和边框*/
    }
    .register label:nth-child(3) .input {
        margin-bottom: 0;                                   /*设置下外边距*/
        border-bottom: 1px solid #d5d5d5;                   /*设置下边框*/
    }
    .register .registerSub {
        margin-top: 18px;                                   /*设置上外边距*/
        width: 300px;                                       /*设置宽度*/
        height: 41px;                                       /*设置高度*/
        line-height: 41px;                                  /*设置行高*/
        background-color: #0f88eb;                          /*设置背景颜色*/
        color: #fff;                                        /*设置文字颜色*/
        font-size: 15px;                                    /*设置文字大小*/
        border: none;                                       /*设置无边框*/
        border-radius: 3px;                                 /*设置圆角边框*/
        text-align: center;                                 /*设置文本水平居中对齐*/
        cursor: pointer;                                    /*设置鼠标形状*/
    }
    .registerSub i {
        margin-left: 5px;                                   /*设置左外边距*/
        font-size: 20px;                                    /*设置文字大小*/
    }
</style>
```

（4）在 store/index.js 文件中定义 mutation。在注册过程中，首先需要判断用户输入的用户名是否已存在。如果用户名不存在，那么我们将注册用户信息存储在 localStorage 中。代码如下：

```js
import { createStore } from 'vuex'
export default createStore({
    mutations: {
        register(state, user) {
            //判断用户名是否已存在
            let isExists = false;
            for(let i = 0; i < localStorage.length; i++)
                if(localStorage.key(i).indexOf('user') !== -1)
                    if(JSON.parse(localStorage.getItem(localStorage.key(i))).name === user.name) {
                        isExists = true;
                        break;
                    }
            if(isExists === false) {
                //将注册用户信息存储在本地
                localStorage.setItem('user'+localStorage.length, JSON.stringify(user));
                sessionStorage.setItem('register','1');              //注册成功
            } else {
                sessionStorage.setItem('register','0');
            }
        }
    }
});
```

7.5.3 用户登录组件设计

用户一旦注册成功，就可以进行登录操作。用户单击"登录"超链接，即可进入用户登录页面，在登录过程中需要输入用户名和密码。登录页面的效果如图 7.10 所示。

用户登录组件的实现过程如下。

（1）在 components 文件夹下创建用户登录组件 UserLogin.vue。在<template>标签中定义用户登录表单，并将表单元素和定义的数据进行绑定。当用户单击"登录"按钮时，将调用 toLog()方法来处理登录请求。代码如下：

图 7.10　在登录页面输入用户名和密码的效果

```html
<template>
    <div class="login">
        <form>
            <label for="name">
                <input type="text" class="input" id="name" v-model="name" placeholder="用户名">
                <em v-show="isShowName">*用户名不正确</em>
            </label>
            <label for="pwd">
                <input type="password" class="input" id="pwd" v-model="pwd" placeholder="密码">
                <em v-show="isShowPwd">*密码不正确</em>
            </label>
            <button type="button" class="loginSub" @click="toLog">
                登　录
                <i class="fa fa-spinner fa-spin" v-show="icon"></i>
            </button>
        </form>
    </div>
</template>
```

（2）在<script>标签中引入 mapMutations()辅助函数，以实现组件中的方法和 store 中的 mutation 之间

的映射。然后定义数据和方法。toReg()方法会检查 sessionStorage 中 key 为 login 的值,以此来判断用户是否成功登录。如果用户登录成功,该方法将页面跳转到通讯录页面。代码如下:

```
<script>
    import { mapMutations } from 'vuex';//引入 mapMutations
    export default {
        data() {
            return {
                name: '',                           //用户名
                pwd: '',                            //密码
                icon: false,                        //控制是否显示旋转图标
                isShowName: false,                  //控制是否显示用户名提示文字
                isShowPwd: false                    //控制是否显示密码提示文字
            };
        },
        methods: {
            ...mapMutations(['login']),
            toLog() {
                this.icon = true;
                this.isShowName = false;
                this.isShowPwd = false;
                setTimeout(() => {
                    this.login({
                        name: this.name,
                        pwd: this.pwd
                    });
                    //判断用户登录
                    if(sessionStorage.getItem('login') === '0'){
                        this.isShowName = true;
                    } else if(sessionStorage.getItem('login') === '1'){
                        this.isShowPwd = true;
                    } else if(sessionStorage.getItem('login') === '2') {
                        this.$router.push({path: '/contacts'});
                    }
                    this.icon = false;
                }, 1000);
            }
        }
    }
</script>
```

(3) 在<style>标签中为用户登录组件中的元素设置样式,代码如下:

```
<style scoped>
    .login {
        width: 994px;                               /*设置宽度*/
        margin: 0 auto;                             /*设置外边距*/
    }
    .login label {
        padding-left: 347px;                        /*设置左内边距*/
        font-size: 15px;                            /*设置文字大小*/
        display: block;                             /*设置元素为块级元素*/
        text-align: left;                           /*设置文本水平左对齐*/
    }
    .register label em {
        margin-left: 5px;                           /*设置左外边距*/
    }
    .login .input {
        width: 300px;                               /*设置宽度*/
        height: 47px;                               /*设置高度*/
        line-height: 47px;                          /*设置行高*/
        margin-bottom: 20px;                        /*设置下外边距*/
```

```css
        border: 1px solid #d5d5d5;                          /*设置边框*/
        padding: 0 8px;                                      /*设置内边距*/
        background-color: #fff;                              /*设置背景颜色*/
        box-sizing: border-box;                              /*宽度和高度包括内容、内边距和边框*/
    }
    .login label:not(:first-child) .input{
        margin-bottom: 0;                                    /*设置下外边距*/
        border-bottom: 1px solid #d5d5d5;                    /*设置下边框*/
    }
    .login .loginSub {
        margin-top: 18px;                                    /*设置上外边距*/
        width: 300px;                                        /*设置宽度*/
        height: 41px;                                        /*设置高度*/
        line-height: 41px;                                   /*设置行高*/
        background-color: #0f88eb;                           /*设置背景颜色*/
        color: #fff;                                         /*设置文字颜色*/
        font-size: 15px;                                     /*设置文字大小*/
        border: none;                                        /*设置无边框*/
        border-radius: 3px;                                  /*设置圆角边框*/
        text-align: center;                                  /*设置文本水平居中对齐*/
        cursor: pointer;                                     /*设置鼠标形状*/
    }
    .loginSub i {
        margin-left: 5px;                                    /*设置左外边距*/
        font-size: 20px;                                     /*设置文字大小*/
    }
</style>
```

（4）在 store/index.js 文件中定义 state 和 mutation。首先，定义 state 中的 user 对象，用于存储当前登录的用户信息。然后，在 mutations 中定义一个 login 方法，该方法在执行登录操作时会进行用户名和密码的验证，并根据验证结果设置 sessionStorage 中的登录状态。如果验证通过，该方法会将用户信息存储在 sessionStorage 中，以便后续使用。代码如下：

```js
import { createStore } from 'vuex'
export default createStore({
    state: {
        user: {}                                             //当前登录用户
    },
    mutations: {
        login(state, user) {
            let currentUser = '';
            let isName = false;
            for(var i = 0; i < localStorage.length; i++)
                if(localStorage.key(i).indexOf('user') !== -1)     //检查键名是否包含 user
                    if(JSON.parse(localStorage.getItem(localStorage.key(i))).name === user.name) {
                        currentUser = JSON.parse(localStorage.getItem(localStorage.key(i)));
                        isName = true;
                        break;
                    }
            if(isName === false) {                                  //判断用户名是否正确
                sessionStorage.setItem('login','0');
            } else if(user.pwd !== currentUser.pwd) {               //判断密码是否正确
                sessionStorage.setItem('login','1');
            } else {
                sessionStorage.setItem('login','2');                //登录成功
                sessionStorage.setItem('user', JSON.stringify(currentUser));  //保存登录用户信息
                sessionStorage.setItem('userId',localStorage.key(i));         //保存登录用户的 key
            }
        }
    }
});
```

7.6 通讯录页面设计

用户登录成功后会跳转到通讯录页面。通讯录页面的左侧是导航栏，右侧是通讯录列表页（包含搜索框、通讯录联系人分类、联系人列表）。通讯录页面效果如图 7.11 所示。

图 7.11 通讯录页面效果

7.6.1 通讯录页面组件设计

通讯录页面组件是一级路由组件。该组件中主要包括左侧导航栏超链接、"退出登录"文本，以及渲染二级路由组件的<router-view>，其实现过程如下。

（1）在 components 文件夹下创建通讯录页面组件 ContactHome.vue。在<template>标签中：首先分别定义通讯录名称、登录用户名；然后使用<router-link>定义"通讯录""添加联系人"和"个人中心"超链接；接着在<div>标签中定义"退出登录"文本，单击该文本调用 out()方法；最后使用<router-view>渲染二级路由组件，并将通讯录列表和当前登录用户信息作为 Prop 属性进行传递。代码如下：

```
<template>
    <div id="mailList">
        <div class="head">
            <div class="logo">畅联</div>
            <div class="welcome">
                <span class="name">{{user.name}}</span>，你好
                <div class="img">
                    <img src="@/assets/img/contact.gif"/>
                </div>
            </div>
        </div>
        <div class="nav">
            <ul>
                <router-link to="/contacts" custom v-slot="{navigate, isExactActive}">
                    <li :class="[isExactActive && 'router-link-exact-active']" @click="navigate">
                        <i class="fa fa-address-book-o"></i> 通讯录
```

```html
            </li>
        </router-link>
        <router-link to="/contacts/add" custom v-slot="{navigate, isExactActive}">
            <li :class="[isExactActive && 'router-link-exact-active']" @click="navigate">
                <i class="fa fa-reorder"></i> 添加联系人
            </li>
        </router-link>
        <router-link to="/contacts/user" custom v-slot="{navigate, isExactActive}">
            <li :class="[isExactActive && 'router-link-exact-active']" @click="navigate">
                <i class="fa fa-user-o"></i> 个人中心
            </li>
        </router-link>
    </ul>
    <div class="out" @click="out">
        <i class="fa fa-cog"></i> 退出登录
    </div>
</div>
<div class="main">
    <router-view :items="items" :user="user" @remove="removeItem"
        @add="addItem" @change="changeItem" @changeUser="changeUser"
    ></router-view>
</div>
</div>
</template>
```

（2）在\<script\>标签中引入 mapState()和 mapMutations()辅助函数，以实现组件中的计算属性、方法和 store 中的 state、mutation 之间的映射。然后定义数据、计算属性和方法。其中：addItem()方法用于调用 mutation 中的 contactsAdd()函数，执行添加联系人信息的操作；removeItem()方法用于调用 mutation 中的 contactsRemove()函数，执行删除联系人信息的操作；changeItem()方法用于调用 mutation 中的 contactsChange()函数，执行修改联系人信息的操作；changeUser()方法用于调用 mutation 中的 userChange()函数，执行修改登录用户信息的操作；out()方法用于退出登录并跳转到登录页面。代码如下：

```javascript
<script>
    import { mapMutations, mapState } from 'vuex'      //引入 mapMutations 和 mapState
    export default {
        data() {
            return {
                tag: 1                                  //激活导航选项样式
            };
        },
        computed:   mapState({
            user: state => state.user,
            items: state => state.items
        }),
        beforeCreate() {
            if(!sessionStorage.getItem('login')){
                alert('请先登录！');
                this.$router.replace('/home/login');    //跳转到登录页面
            }
            this.$store.commit('contactsInit');
        },
        methods: {
            ...mapMutations([
                'logout',
                'contactsAdd',
                'contactsRemove',
                'contactsChange',
                'userChange'
            ]),
            addItem(item) {
```

```
                this.contactsAdd(item);                    //添加联系人信息
            },
            removeItem(id) {
                this.contactsRemove(id);                   //删除联系人信息
            },
            changeItem(obj) {
                this.contactsChange(obj);                  //修改联系人信息
            },
            changeUser(obj) {
                this.userChange(obj);                      //修改登录用户信息
            },
            out() {
                this.logout();                             //退出登录
                this.$router.replace('/home/login');       //跳转到登录页面
            }
        }
    }
</script>
```

（3）在<style>标签中为通讯录页面组件中的元素设置样式，代码如下：

```
<style scoped>
    body, ul, h1, h2, p {
        padding: 0;                                        /*设置内边距*/
        margin: 0;                                         /*设置外边距*/
    }
    html {
        font-size: 10px;                                   /*设置文字大小*/
    }
    html, body, #mailList {
        height: 100%;                                      /*设置高度*/
        color: #555;                                       /*设置文字颜色*/
    }
    body {
        overflow-x: hidden;                                /*对内容的左/右边缘进行裁剪但不滚动*/
    }
    #mailList {
        width: 100%;                                       /*设置宽度*/
    }
    .head {
        width: 100%;                                       /*设置宽度*/
        height: 90px;                                      /*设置高度*/
        line-height: 90px;                                 /*设置行高*/
        position: fixed;                                   /*设置相对于浏览器进行定位*/
        top: 0;                                            /*设置到浏览器顶端距离*/
        background-color: #fff;                            /*设置背景颜色*/
        display: flex;                                     /*设置弹性布局*/
        justify-content: space-between;                    /*设置两端对齐*/
        z-index: 5;                                        /*设置元素的堆叠顺序*/
        background-size: cover;                            /*设置背景图像尺寸*/
    }
    .head .logo {
        font-size: 50px;                                   /*设置文字大小*/
        padding-left: 30px;                                /*设置左内边距*/
        color: #0f88eb;                                    /*设置文字颜色*/
    }
    .head .welcome {
        font-size: 20px;                                   /*设置文字大小*/
        display: inline-block;                             /*设置元素为行内块元素*/
        margin-right: 30px;                                /*设置右外边距*/
    }
    .head .img {
```

```css
        display: inline-block;            /*设置元素为行内块元素*/
        width: 60px;                      /*设置宽度*/
        height: 60px;                     /*设置高度*/
        border-radius: 35px;              /*设置圆角边框*/
        vertical-align: middle;           /*设置垂直居中对齐*/
        margin-left: 10px;                /*设置左外边距*/
        background-color: #eee;           /*设置背景颜色*/
        overflow: hidden;                 /*设置溢出内容隐藏*/
    }
    .head img {
        width: 100%;                      /*设置宽度*/
        height: 100%;                     /*设置高度*/
    }
    .nav {
        width: 300px;                     /*设置宽度*/
        height: 90%;                      /*设置高度*/
        position: fixed;                  /*设置相对于浏览器进行定位*/
        left: 0;                          /*设置到浏览器左端距离*/
        top: 90px;                        /*设置到浏览器顶端距离*/
        z-index: 5;                       /*设置元素的堆叠顺序*/
        background-color: #eee;           /*设置背景颜色*/
        padding: 3px;                     /*设置内边距*/
        box-sizing: border-box;           /*宽度和高度包括内容、内边距和边框*/
    }
    .nav li {
        height: 50px;                     /*设置高度*/
        line-height: 50px;                /*设置行高*/
        padding-left: 30px;               /*设置左内边距*/
        margin-bottom: 2px;               /*设置下外边距*/
        box-sizing: border-box;           /*宽度和高度包括内容、内边距和边框*/
        font-size: 17px;                  /*设置文字大小*/
        border-top: none;                 /*设置无上边框*/
        cursor: pointer;                  /*设置鼠标形状*/
        background-color: #fff;           /*设置背景颜色*/
        list-style: none;                 /*设置列表无样式*/
    }
    .nav li i {
        margin-right: 8px;                /*设置右外边距*/
    }
    .main {
        width: 100%;                      /*设置宽度*/
        height: 100%;                     /*设置高度*/
        background-color: #f7fafc;        /*设置背景颜色*/
        position: absolute;               /*设置绝对定位*/
        padding-left: 300px;              /*设置左内边距*/
        box-sizing: border-box;           /*宽度和高度包括内容、内边距和边框*/
        z-index: 1;                       /*设置元素的堆叠顺序*/
        overflow-x: hidden;               /*对内容的左/右边缘进行裁剪但不滚动*/
    }
    .router-link-exact-active{
        background-color: #0f88eb !important;  /*设置背景颜色*/
        color: #fff;                      /*设置文字颜色*/
    }
    .name {
        display: inline-block;            /*设置元素为行内块元素*/
        max-width: 118px;                 /*设置最大宽度*/
        white-space: nowrap;              /*设置段落中的文本不进行换行*/
        text-overflow: ellipsis;          /*溢出内容使用省略号*/
        overflow: hidden;                 /*设置溢出内容隐藏*/
        vertical-align: middle;           /*设置垂直居中对齐*/
    }
    .out {
```

```
            position: absolute;                              /*设置绝对定位*/
            bottom: 41px;                                    /*设置到父元素底端距离*/
            right: 16px;                                     /*设置到父元素右端距离*/
            cursor: pointer;                                 /*设置鼠标形状*/
            font-size: 16px;                                 /*设置文字大小*/
        }
</style>
```

（4）在 store/index.js 文件中定义 state 和 mutation。其中，items 用于存储通讯录列表。在 mutation 中，logout()方法用于退出登录，contactsInit()方法用于初始化定义的 state，contactsAdd()方法用于添加联系人信息，contactsRemove()方法用于删除联系人信息，contactsChange()方法用于修改联系人信息，userChange()方法用于修改当前登录用户信息。代码如下：

```
import { createStore } from 'vuex'
let contactId = 0;                                           //联系人 id
export default createStore({
    state: {
        //联系人列表
        items: localStorage.getItem('items') ? JSON.parse(localStorage.getItem('items')) : [
            {name: '张三', tel: 84978981, type: "朋友"},
            {name: 'Alice', tel: 84978981, type: "朋友"},
            {name: '李四', tel: 84978981, type: "亲人"},
            {name: 'Jack', tel: 84978981, type: "朋友"},
            {name: '王五', tel: 84978981, type: "同事"},
            {name: 'Rose', tel: 84978981, type: "亲人"},
            {name: '赵六', tel: 84978981, type: "同学"},
            {name: 'Smith', tel: 84978981, type: "同事"},
            {name: 'Kelly', tel: 84978981, type: "亲人"},
            {name: 'Tom', tel: 84978981, type: "同学"},
            {name: '大姨', tel: 84978981, type: "亲人"},
            {name: 'Robert', tel: 84978981, type: "同学"},
            {name: '二舅', tel: 84978981, type: "亲人"},
            {name: 'Henry', tel: 84978981, type: "同事"}
        ]
    },
    mutations: {
        logout() {                                           //退出登录
            sessionStorage.setItem('register', '0');
            sessionStorage.setItem('login', '0');
        },
        contactsInit(state) {                                //初始化 state
            state.items.forEach((item)=>{
                item.id = contactId++;
                item.imgSrc = require('@/assets/img/userImg.png');
            });
            state.user = sessionStorage.getItem('user') ? JSON.parse(sessionStorage.getItem('user')) : {};
        },
        contactsAdd(state, contacts) {                       //添加联系人
            contacts.id = contactId++;
            contacts.imgSrc = require('@/assets/img/userImg.png');
            state.items.push(contacts);
            localStorage.setItem('items', JSON.stringify(state.items));
        },
        contactsRemove(state, contactsId) {                  //删除联系人
            state.items = state.items.filter(function(item) {
                return item.id !== contactsId;
            });
            localStorage.setItem('items', JSON.stringify(state.items));
        },
        contactsChange(state, contacts) {                    //修改联系人信息
            for(let key in state.items){
```

```
            if(state.items[key].id === contacts.id) {
                state.items[key].name = contacts.name;
                state.items[key].tel = contacts.tel;
            }
        }
        localStorage.setItem('items', JSON.stringify(state.items));
    },
    userChange(state, user) {                              //修改当前登录用户信息
        state.user = user;
        sessionStorage.setItem('user', JSON.stringify(user));  //保存登录用户信息
        //保存注册用户信息
        localStorage.setItem(sessionStorage.getItem('userId'), JSON.stringify(user));
    }
}
});
```

7.6.2 通讯录列表组件设计

通讯录列表组件主要包括搜索框、通讯录联系人分类和联系人列表,其页面效果如图 7.12 所示。

图 7.12 通讯录列表页面效果

通讯录列表组件的实现过程如下。

(1) 在 components 文件夹下创建通讯录列表组件 ShowContacts.vue。在<template>标签中,分别定义搜索框、联系人总数和表示联系人分类的单选按钮,选中不同的单选按钮可以选择不同的分类。在单选按钮下方调用分页组件 DataPage,将筛选后的联系人列表和每页显示的联系人个数作为 Prop 属性进行传递。代码如下:

```
<template>
    <div id="search">
        <div class="search">
            <input type="text" placeholder="请输入搜索联系人" v-model="keywords">
            <span>
                <i class="fa fa-search"></i>
            </span>
        </div>
        <p class="all">共有 {{items.length}} 个联系人</p>
        <div class="select">
            <input type="radio" name="contacts" checked @click="selectType=1">所有联系人
```

```
            <input type="radio" name="contacts" @click="selectType=2">亲人
            <input type="radio" name="contacts" @click="selectType=3">朋友
            <input type="radio" name="contacts" @click="selectType=4">同事
            <input type="radio" name="contacts" @click="selectType=5">同学
        </div>
        <div class="ul" v-if="newItems.length">
            <DataPage
                :items="newItems"
                :pageSize="12"
                @remove="remove"
                @change="change"
            />
        </div>
        <p class="none" v-else>
            没有联系人
        </p>
    </div>
</template>
```

（2）在<script>标签中引入分页组件 DataPage，然后定义数据、方法和计算属性。其中，remove()方法用于触发自定义事件 remove，change()方法用于触发自定义事件 change，newItems 计算属性用于根据分类和搜索关键字获取联系人列表中的指定联系人。代码如下：

```
<script>
    import DataPage from './DataPage'                          //引入分页组件
    export default {
        props: ["items"],
        data() {
            return {
                keywords: '',                                   //搜索关键字
                selectType: 1,                                  //联系人分类
            }
        },
        methods: {
            remove(id) {                                        //删除联系人
                this.$emit('remove', id);
            },
            change(obj) {                                       //修改联系人
                this.$emit('change', obj);
            }
        },
        computed: {
            newItems: function () {
                if(this.items.length===0)
                    return '';
                if(this.keywords) {
                    //根据分类获取联系人姓名中包含搜索关键字的联系人
                    switch (this.selectType) {
                        case 1:
                            return this.items.filter((item)=> {
                                return (item.name.indexOf(this.keywords) !== -1);
                            });
                        case 2:
                            return this.items.filter((item)=> {
                                return (item.name.indexOf(this.keywords) !== -1 && item.type === '亲人');
                            });
                        case 3:
                            return this.items.filter((item)=> {
                                return (item.name.indexOf(this.keywords) !== -1 && item.type === '朋友');
                            });
                        case 4:
                            return this.items.filter((item)=> {
```

```
                        return (item.name.indexOf(this.keywords) !== -1 && item.type === '同事');
                    });
                default:
                    return this.items.filter((item)=> {
                        return (item.name.indexOf(this.keywords) !== -1 && item.type === '同学');
                    });
                }
            } else {
                //根据分类获取联系人列表
                switch (this.selectType) {
                    case 1:
                        return this.items;
                    case 2:
                        return this.items.filter(function (item) {
                            return item.type === '亲人';
                        });
                    case 3:
                        return this.items.filter(function (item) {
                            return item.type === '朋友';
                        });
                    case 4:
                        return this.items.filter(function (item) {
                            return item.type === '同事';
                        });
                    default:
                        return this.items.filter(function (item) {
                            return item.type === '同学';
                        });
                }
            }
        }
    },
    components: {
        DataPage
    }
}
</script>
```

（3）在\<style\>标签中为通讯录列表组件中的元素设置样式，代码如下：

```
<style scoped>
    #search {
        padding-top: 90px;                              /*设置上内边距*/
    }
    .search {
        padding-left: 181px;                            /*设置左内边距*/
        margin-top: 39px;                               /*设置上外边距*/
    }
    .search input {
        width: 640px;                                   /*设置宽度*/
        height: 39px;                                   /*设置高度*/
        border: 2px solid #eee;                         /*设置边框*/
        border-right: none;                             /*设置无右边框*/
        border-top-left-radius: 5px;                    /*设置左上圆角边框*/
        border-bottom-left-radius: 5px;                 /*设置左下圆角边框*/
        padding-left: 10px;                             /*设置左内边距*/
        vertical-align: middle;                         /*设置垂直居中对齐*/
        font-size: 16px;                                /*设置文字大小*/
    }
    input:focus, button:focus {
        outline: none;                                  /*设置元素无轮廓*/
    }
    input[type="radio"] {
```

```css
    cursor: pointer;                              /*设置鼠标形状*/
}
.search span {
    background-color: #fff;                       /*设置背景颜色*/
    color: #555;                                  /*设置文字颜色*/
    font-size: 16px;                              /*设置文字大小*/
    padding: 10px 14px;                           /*设置内边距*/
    border: 2px solid #eee;                       /*设置边框*/
    border-top-right-radius: 5px;                 /*设置右上圆角边框*/
    border-bottom-right-radius: 5px;              /*设置右下圆角边框*/
    border-left: none;                            /*设置无左边框*/
    vertical-align: middle;                       /*设置垂直居中对齐*/
}
.all {
    font-size: 13px;                              /*设置文字大小*/
    margin: 10px 0;                               /*设置外边距*/
    padding-left: 191px;                          /*设置左内边距*/
}
.select {
    padding-left: 181px;                          /*设置左内边距*/
    font-size: 14px;                              /*设置文字大小*/
}
.main .ul {
    width: 1000px;                                /*设置宽度*/
    height: auto;                                 /*设置高度*/
    margin-top: 39px;                             /*设置上外边距*/
    margin-left: 132px;                           /*设置左外边距*/
    padding-left: 50px;                           /*设置左内边距*/
    box-sizing: border-box;                       /*宽度和高度包括内容、内边距和边框*/
}
.none {
    margin-top: 122px;                            /*设置上外边距*/
    font-size: 33px;                              /*设置文字大小*/
    padding-left: 340px;                          /*设置左内边距*/
}
</style>
```

7.6.3 分页组件设计

当联系人比较多时,为了查看方便,我们需要为联系人列表设计分页显示功能。为此,我们需要创建分页组件,作为通讯录列表组件的子组件。在分页组件中,用户可以查看当前页数和总页数,并通过单击"<"或">"按钮来切换至上一页或下一页中的联系人列表。分页组件的页面效果如图 7.13 所示。

分页组件的实现过程如下。

（1）在 components 文件夹下创建分页组件 DataPage.vue。在<template>

图 7.13 分页组件的页面效果

标签中,首先使用 v-for 指令遍历联系人组件 ContactItem,将分页数据作为 Prop 属性进行传递,然后定义用于实现分页跳转的 "<" 和 ">" 按钮,用户单击 "<" 按钮时会调用 prevPage()方法,单击 ">" 按钮时会调用 nextPage()方法。代码如下：

```vue
<template>
    <div class="page">
        <ContactItem v-for="item in pageItems" :key="item"
                     :item="item" @change="change" @remove="remove">
        </ContactItem>
        <div class="page_control">
            <button @click="prevPage" :disabled = "currentPage === 1">&lt;</button>
            <span>{{currentPage}}/{{totalPages}}</span>
```

```html
        <button @click="nextPage" :disabled = "currentPage === totalPages">&gt;</button>
      </div>
    </div>
</template>
```

（2）在<script>标签中引入联系人组件 ContactItem，然后定义数据、方法和计算属性。其中，remove()方法用于删除联系人信息，change()方法用于修改联系人信息，prevPage()方法用于将当前页减 1，nextPage()方法用于将当前页码加 1，totalPages 计算属性用于获取总页数，pageItems 计算属性用于获取分页数据。代码如下：

```javascript
<script>
import ContactItem from "@/components/ContactItem.vue";
export default {
    props: ['items','pageSize'],
    data() {
        return {
            currentPage: 1                                      //当前页数
        };
    },
    components: {
        ContactItem
    },
    methods: {
        remove(id) {                                            //删除联系人
            this.$emit('remove', id);
        },
        change(obj) {                                           //修改联系人
            this.$emit('change', obj);
        },
        prevPage() {
            if(this.currentPage > 1) this.currentPage -= 1;     //当前页减 1
        },
        nextPage() {
            if(this.currentPage < this.totalPages) this.currentPage += 1;   //当前页加 1
        }
    },
    computed: {
        totalPages(){
            return Math.ceil(this.items.length / this.pageSize);    //获取总页数
        },
        pageItems(){
            let start = (this.currentPage - 1) * this.pageSize;
            let end = start + this.pageSize;
            return this.items.slice(start, end);
        }
    }
}
</script>
```

（3）在<style>标签中为分页组件中的元素设置样式，代码如下：

```css
<style scoped>
    .page .page_control{
        margin-left: 240px;                 /*设置左外边距*/
        margin-bottom: 10px;                /*设置下外边距*/
    }
    .page .page_control span{
        margin: 0 10px;                     /*设置外边距*/
    }
    .page .page_control button{
        width: 60px;                        /*设置宽度*/
        height: 30px;                       /*设置高度*/
    }
</style>
```

7.6.4 联系人组件设计

联系人列表由多个联系人构成。为了代码的可读性和可维护性，需要将单个联系人的页面结构单独定义为一个联系人组件。联系人组件是分页组件的子组件。在该组件中除了显示联系人信息，还提供了修改和删除联系人的功能。联系人组件的页面效果如图 7.14 所示。单击联系人信息右侧的编辑图标，页面效果如图 7.15 所示。

联系人组件的实现过程如下。

（1）在 components 文件夹下创建联系人组件 ContactItem.vue。在 <template> 标签中定义 <div> 标签，用于展示联系人信息，包括联系人头像、联系人姓名和联系人电话号码。右侧配有图标按钮，用于表示编辑和删除操作。当用户单击编辑图标按钮时，会显示编辑联系人的页面，在该页面中可以对联系人信息进行编辑或取消编辑。代码如下：

图 7.14　联系人组件的页面效果

图 7.15　编辑联系人的页面效果

```html
<template>
  <div class="li">
    <div v-if="show">
      <div class="text">
        <img :src="item.imgSrc"/>
        <div>
          <h1 class="name">{{item.name}}</h1>
          <p class="tel">{{item.tel}}</p>
        </div>
      </div>
      <button class="edit" @click="show=false"><i class="fa fa-edit"></i></button>
      <button class="delete" @click="toRemove(item.id)"><i class="fa fa-trash-o"></i></button>
    </div>
    <div v-else>
      <div class="text">
        <img :src="item.imgSrc"/>
        <div>
          <input class="name" type="text" v-model="name"/>
          <input class="tel" type="text" v-model="tel"/>
        </div>
      </div>
      <button class="save" @click="save()"><i class="fa fa-check"></i></button>
      <button class="return" @click="notDo()"><i class="fa fa-times"></i></button>
    </div>
  </div>
</template>
```

（2）在 <script> 标签中定义数据和方法。当用户单击删除图标按钮时，将调用 toRemove() 方法，该方法用于删除联系人信息。用户对联系人信息进行重新编辑后：如果用户单击对号图标会调用 save() 方法以保存修改后的联系人信息；如果用户单击叉号图标，则会调用 notDo() 方法来取消编辑。代码如下：

```html
<script>
export default {
  props: ['item'],
  data() {
    return {
      show: true,                        //控制显示内容
      name: this.item.name,              //联系人姓名
      tel: this.item.tel,                //联系人电话号码
    };
  },
  methods: {
```

```
        toRemove(id) {                              //删除联系人信息
            this.$emit('remove', id);
        },
        save() {                                    //保存修改后的联系人信息
            this.show = true;
            this.$emit('change', {
                id: this.item.id,
                name: this.name,
                tel: this.tel
            });
        },
        notDo() {                                   //不修改
            this.show = true;
            this.name = this.item.name;
            this.tel = this.item.tel;
        }
    }
}
</script>
```

（3）在<style>标签中，为联系人组件中的元素设置样式，代码如下：

```
<style scoped>
    .li {
        width: 220px;                               /*设置宽度*/
        margin-right: 48px;                         /*设置右外边距*/
        list-style: none;                           /*设置列表无样式*/
        height: 66px;                               /*设置高度*/
        margin-bottom: 25px;                        /*设置下外边距*/
        display: inline-block;                      /*设置元素为行内块元素*/
    }
    .li .text {
        width: 163px;                               /*设置宽度*/
        padding: 3px 0;                             /*设置内边距*/
        display: inline-block;                      /*设置元素为行内块元素*/
    }
    .text > div {
        margin-top: 3px;                            /*设置上外边距*/
        display: inline-block;                      /*设置元素为行内块元素*/
        vertical-align: middle;                     /*设置垂直居中对齐*/
    }
    .text img {
        width: 50px;                                /*设置宽度*/
        height: 50px;                               /*设置高度*/
        margin-right: 2px;                          /*设置右外边距*/
        vertical-align: middle;                     /*设置垂直居中对齐*/
    }
    .text input {
        display: block;                             /*设置元素为块级元素*/
        border: none;                               /*设置无边框*/
    }
    .text .name {
        width: 105px;                               /*设置宽度*/
        margin-bottom: 3px;                         /*设置下外边距*/
        font-size: 18px;                            /*设置文字大小*/
        margin-top: 0;                              /*设置上外边距*/
    }
    .text .tel {
        width: 105px;                               /*设置宽度*/
        font-size: 15px;                            /*设置文字大小*/
        margin: 0;                                  /*设置外边距*/
    }
    .li button {
```

```
            height: 32px;                                /*设置高度*/
            padding: 0;                                  /*设置内边距*/
            margin-top: 11px;                            /*设置上外边距*/
            border: none;                                /*设置无边框*/
            color: #0F88EC;                              /*设置文字颜色*/
            background: none;                            /*设置无背景*/
            border-radius: 4px;                          /*设置圆角边框*/
            cursor: pointer;                             /*设置鼠标形状*/
            display: none;                               /*设置元素隐藏*/
        }
        .li button.delete {
            color: #D81111;                              /*设置文字颜色*/
        }
        .save, .edit {
            margin-right: 3px;                           /*设置右外边距*/
        }
        .li button:focus, .text input:focus {
            outline: none;                               /*设置元素无轮廓*/
        }
        .li i {
            font-size: 20px;                             /*设置文字大小*/
        }
        .li:hover {
            border: 1px solid #009eef;                   /*设置边框*/
            border-radius: 5px;                          /*设置圆角边框*/
        }
        .li:hover button {
            display: inline-block;                       /*设置元素为行内块元素*/
        }
</style>
```

7.7 添加联系人组件设计

单击通讯录页面左侧导航栏中的"添加联系人"超链接，页面右侧将展示添加联系人的页面，如图7.16所示。在添加联系人时，用户需要输入联系人的姓名、电话号码，以及联系人与当前登录用户的关系，如图7.17所示。输入完成后，单击"添加"按钮，页面将跳转回通讯录页面，并在通讯录列表的最后一页显示新添加的联系人信息，如图7.18所示。

图7.16 添加联系人页面

图 7.17 输入联系人信息

图 7.18 显示新添加的联系人

添加联系人组件的实现过程如下。

（1）在 components 文件夹下创建添加联系人组件 AddCon.vue。在<template>标签中定义 4 个<label>标签，在这 4 个标签中分别定义用于输入联系人姓名的文本框、用于输入联系人电话号码的文本框、用于选择联系人和当前登录用户关系的下拉菜单和一个"添加"按钮。当这两个文本框获得焦点时，将调用 judge()方法；当用户单击"添加"按钮时，将调用 add()方法。代码如下：

```
<template>
    <div>
        <h2>添加联系人</h2>
        <div class="form">
            <label for="name">
                <span>姓名</span>
                <input type="text" id="name" @focus="judge(1)" v-model="name" :class="{'error': isShowName}"
                    placeholder="请输入姓名" />
                <em v-show="isShowName">*姓名只能由字母、数字或下画线组成</em>
            </label>
```

```html
            <label for="tel">
                <span>电话号码</span>
                <input type="text" id="tel" @focus="judge(2)" v-model="tel" :class="{'error': isShowTel}"
                    placeholder="请输入电话号码">
                <em v-show="isShowTel">*电话号码只能由数字组成</em>
            </label>
            <label>
                <span>与我的关系</span>
                <select v-model="type">
                    <option selected>亲人</option>
                    <option>朋友</option>
                    <option>同事</option>
                    <option>同学</option>
                </select>
            </label>
            <label>
                <span></span>
                <button class="normal" :class="{'btn': save}" @click="add" :disabled="save">添　加</button>
            </label>
        </div>
    </div>
</template>
```

（2）在\<script\>标签中定义数据、方法、计算属性和监听属性。其中：judge()方法会根据传递的参数判断姓名或电话号码是否符合要求；add()方法会通过\$emit()方法触发自定义事件add，将联系人信息作为参数进行传递，并跳转到通讯录页面；计算属性save用于控制"添加"按钮是否可用。watch选项用于对用户输入的联系人姓名和电话号码进行监听，如果用户输入的内容不符合要求，就显示相应的提示文字。代码如下：

```js
<script>
export default {
    data() {
        return {
            name: '',                                    //联系人姓名
            tel: '',                                     //联系人电话号码
            type: '亲人',                                 //联系人类型
            isShowName: false,                           //是否显示联系人姓名提示信息
            isShowTel: false                             //是否显示联系人电话号码提示信息
        }
    },
    methods: {
        judge(n){
            if(n === 1){
                //判断姓名是否符合要求
                this.isShowName = this.name === '' || /\W/g.test(this.name);
            } else {
                //判断电话号码是否符合要求
                this.isShowTel = this.tel==='' || /\D/g.test(this.tel);
            }
        },
        add() {                                          //添加联系人
            if(this.name && this.tel) {
                this.$emit('add', {
                    name: this.name,
                    tel: this.tel,
                    type: this.type,
                });
                this.name = '';
                this.tel='';
                this.$router.push('/contacts');          //跳转到通讯录页面
            } else {
                this.isShowName = this.name === '' || /\W/g.test(this.name);
```

```
                    this.isShowTel = this.tel==='' || /\D/g.test(this.tel);
                }
            }
        },
        computed: {
            save() {                                            //控制添加按钮是否可用
                return this.isShowName || this.isShowTel;
            }
        },
        watch: {
            name: function (val){                               //监听联系人姓名
                this.isShowName = /\W/g.test(val);
            },
            tel: function (val){                                //监听联系人电话号码
                this.isShowTel = /\D/g.test(val);
            }
        }
    }
</script>
```

（3）在<style>标签中，为添加联系人组件中的元素设置样式，代码如下：

```
<style scoped>
    .main h2 {
        width: 84%;                                             /*设置宽度*/
        text-align: center;                                     /*设置文本水平居中对齐*/
        padding: 20px;                                          /*设置内边距*/
        margin-top: 138px;                                      /*设置上外边距*/
        font-size: 38px;                                        /*设置文字大小*/
    }
    .form {
        width: 82%;                                             /*设置宽度*/
        padding-top: 20px;                                      /*设置上内边距*/
    }
    .form label {
        display: block;                                         /*设置元素为块级元素*/
        width: 770px;                                           /*设置宽度*/
        height: 42px;                                           /*设置高度*/
        margin-left: 256px;                                     /*设置左外边距*/
        margin-bottom: 15px;                                    /*设置下外边距*/
    }
    .form span {
        display: inline-block;                                  /*设置元素为行内块元素*/
        width: 87px;                                            /*设置宽度*/
        font-size: 16px;                                        /*设置文字大小*/
        text-align: right;                                      /*设置文本水平右对齐*/
        margin-right: 10px;                                     /*设置右外边距*/
    }
    .form input {
        width: 300px;                                           /*设置宽度*/
        height: 40px;                                           /*设置高度*/
        padding-left: 10px;                                     /*设置左内边距*/
        box-sizing: border-box;                                 /*宽度和高度包括内容、内边距和边框*/
        font-size: 16px;                                        /*设置文字大小*/
        border-radius: 5px;                                     /*设置圆角边框*/
        border: 1px solid #ccc;                                 /*设置边框*/
        background-color: #fff;                                 /*设置背景颜色*/
    }
    .form select {
        height: 35px;                                           /*设置高度*/
        width: 150px;                                           /*设置宽度*/
        border: 1px solid #ccc;                                 /*设置边框*/
```

```
        padding-left: 10px;                              /*设置左内边距*/
        box-sizing: border-box;                          /*宽度和高度包括内容、内边距和边框*/
    }
    .form select:focus, .form button:focus {
        outline: none;                                   /*设置元素无轮廓*/
    }
    .form button.normal {
        width: 150px;                                    /*设置宽度*/
        border: none;                                    /*设置无边框*/
        height: 40px;                                    /*设置高度*/
        font-size: 18px;                                 /*设置文字大小*/
        border-radius: 5px;                              /*设置圆角边框*/
        cursor: pointer;                                 /*设置鼠标形状*/
        background-color: #0f88eb;                       /*设置背景颜色*/
        color: #FFFFFF;                                  /*设置文字颜色*/
    }
    .form button.btn {
        background-color: #ccc;                          /*设置背景颜色*/
    }
    input.error {
        border: 2px solid #ff0000;                       /*设置边框*/
    }
    em {
        color: #FF0000;                                  /*设置文字颜色*/
    }
</style>
```

7.8 个人中心组件设计

单击通讯录页面左侧导航栏中的"个人中心"超链接,页面右侧将展示当前登录用户个人信息的页面,效果如图 7.19 所示。单击个人信息界面中的"编辑信息"按钮,用户可以对个人信息进行编辑,包括用户名、电话号码和密码,如图 7.20 所示。编辑信息后,单击"保存"按钮,页面将显示编辑后的个人信息,如图 7.21 所示。

图 7.19 个人信息页面

图 7.20　编辑个人信息

图 7.21　编辑后的个人信息

个人中心组件的实现过程如下。

（1）在 components 文件夹下创建个人中心组件 SaveUser.vue。在<template>标签中定义 5 个<label>标签。其中：第 1 个<label>标签用于定义输入用户名的文本框；第 2 个<label>标签用于定义输入电话号码的文本框；第 3 个<label>标签用于定义输入密码的文本框；第 4 个<label>标签用于定义一个"编辑信息"按钮；第 5 个<label>标签用于定义一个"保存"按钮和一个"返回"按钮。在默认情况下，这 3 个文本框都是不可编辑的状态。当单击"编辑信息"按钮时，将 edit 属性的值设置为 false，这时会隐藏"编辑信息"按钮，并显示"保存"按钮和"返回"按钮，同时 3 个文本框变成可编辑的状态；当单击"保存"按钮时，将调用 change()方法；当单击"返回"按钮时，将调用 notChange()方法。代码如下：

```
<template>
    <div>
        <h2>个人信息</h2>
        <div class="form">
            <label for="name">
                <span>用户名：</span>
```

```html
            <input :class="className.name" type="text" id="name" v-model="name" :disabled='edit'/>
            <em v-show="isShowName">*用户名只能由字母、数字或下画线组成</em>
        </label>
        <label for="tel">
            <span>电话号码：</span>
            <input :class="className.tel" type="text" id="tel" v-model="tel" :disabled='edit'/>
            <em v-show="isShowTel">*电话号码只能由数字组成</em>
        </label>
        <label for="pwd">
            <span>密码：</span>
            <input :class="className.pwd" type="text" id="pwd" v-model="pwd" :disabled='edit'/>
            <em v-show="isShowPwd">*密码不能小于6位</em>
        </label>
        <label v-if="edit">
            <span></span>
            <button class="normal large" @click="edit=false">编辑信息</button>
        </label>
        <label class="editing" v-else>
            <span></span>
            <button :class="{'light': !save, 'dark': save}" @click="change" :disabled="save">保存</button>
            <button class="light" @click="notChange">返回</button>
        </label>
    </div>
  </div>
</template>
```

（2）在`<script>`标签中定义Prop属性、数据、方法、计算属性和监听属性。其中：change()方法通过$emit()方法触发自定义事件changeUser，将编辑后的个人信息作为参数进行传递；notChange()方法将3个文本框的值分别设置为默认显示的值，实现不修改个人信息的目的；计算属性save用于控制"保存"按钮是否可用，计算属性className用于控制3个文本框的显示样式；watch选项用于对用户输入的用户名、电话号码和密码进行监听，如果用户输入的内容不符合要求，就显示相应的提示文字。代码如下：

```javascript
<script>
    export default {
        props: ["user"],
        data() {
            return {
                name: this.user.name,              //用户名
                tel: this.user.tel,                //电话号码
                pwd: this.user.pwd,                //密码
                edit: true,                        //控制输入框样式和是否可用
                isShowName: false,                 //是否显示用户名提示信息
                isShowTel: false,                  //是否显示电话号码提示信息
                isShowPwd: false                   //是否显示密码提示信息
            }
        },
        methods: {
            change() {                             //修改用户信息
                if(this.name && this.tel && this.pwd) {
                    this.edit = true;
                    this.$emit('changeUser', {
                        name: this.name,
                        tel: this.tel,
                        pwd: this.pwd
                    });
                }
            },
            notChange() {                          //不修改
                this.edit = true;
                this.name = this.user.name;
                this.tel = this.user.tel;
```

```
                this.pwd = this.user.pwd;
                this.isShowName = false;
                this.isShowTel = false;
                this.isShowPwd = false;
            }
        },
        computed: {
            //是否禁用保存按钮
            save(){
                return this.isShowName || this.isShowTel || this.isShowPwd;
            },
            className: function() {
                return {
                    name: {
                        only_read: this.edit,                       //是否显示 only_read 类样式
                        error: this.isShowName                      //是否显示 error 类样式
                    },
                    tel: {
                        only_read: this.edit,
                        error: this.isShowTel
                    },
                    pwd: {
                        only_read: this.edit,
                        error: this.isShowPwd
                    }
                };
            }
        },
        watch: {
            name: function (val){                                   //监听用户名
                this.isShowName = /\W/g.test(val) || val === '';
            },
            tel: function (val){                                    //监听用户电话号码
                this.isShowTel = /\D/g.test(val) || val === '';
            },
            pwd: function (val){                                    //监听用户密码
                this.isShowPwd = val.length < 6;
            }
        }
    }
</script>
```

（3）在<style>标签中，为个人中心组件中的元素设置样式，代码如下：

```
<style>
    .main h2 {
        width: 84%;                                                 /*设置宽度*/
        text-align: center;                                         /*设置文本水平居中对齐*/
        padding: 20px;                                              /*设置内边距*/
        margin-top: 138px;                                          /*设置上外边距*/
        font-size: 38px;                                            /*设置文字大小*/
    }
    .form {
        width: 82%;                                                 /*设置宽度*/
        padding-top: 20px;                                          /*设置上内边距*/
    }
    .form label {
        display: block;                                             /*设置元素为块级元素*/
        width: 770px;                                               /*设置宽度*/
        height: 42px;                                               /*设置高度*/
        margin-left: 256px;                                         /*设置左外边距*/
        margin-bottom: 15px;                                        /*设置下外边距*/
    }
```

```css
.form span {
    display: inline-block;              /*设置元素为行内块元素*/
    width: 87px;                        /*设置宽度*/
    font-size: 16px;                    /*设置文字大小*/
    text-align: right;                  /*设置文本水平右对齐*/
    margin-right: 10px;                 /*设置右外边距*/
}
.form input {
    width: 300px;                       /*设置宽度*/
    height: 40px;                       /*设置高度*/
    padding-left: 10px;                 /*设置左内边距*/
    box-sizing: border-box;             /*宽度和高度包括内容、内边距和边框*/
    font-size: 16px;                    /*设置文字大小*/
    border-radius: 5px;                 /*设置圆角边框*/
    border: 1px solid #ccc;             /*设置边框*/
    background-color: #fff;             /*设置背景颜色*/
}
.form select {
    height: 35px;                       /*设置高度*/
    width: 150px;                       /*设置宽度*/
    border: 1px solid #ccc;             /*设置边框*/
    padding-left: 10px;                 /*设置左内边距*/
    box-sizing: border-box;             /*宽度和高度包括内容、内边距和边框*/
}
.form select:focus, .form button:focus {
    outline: none;                      /*设置元素无轮廓*/
}
.form button.normal {
    width: 150px;                       /*设置宽度*/
    border: none;                       /*设置无边框*/
    height: 40px;                       /*设置高度*/
    font-size: 18px;                    /*设置文字大小*/
    border-radius: 5px;                 /*设置圆角边框*/
    cursor: pointer;                    /*设置鼠标形状*/
    background-color: #0f88eb;          /*设置背景颜色*/
    color: #fff;                        /*设置文字颜色*/
}
.form button.btn {
    background-color: #ccc;             /*设置背景颜色*/
}
input.error {
    border: 2px solid #ff0000;          /*设置边框*/
}
em {
    color: #FF0000;                     /*设置文字颜色*/
}
.form input.only_read {
    background-color: #f7fafc;          /*设置背景颜色*/
    border: none;                       /*设置无边框*/
    text-align: center;                 /*设置文本水平居中对齐*/
}
.form button.large {
    width: 300px;                       /*设置宽度*/
}
.editing button {
    width: 147px;                       /*设置宽度*/
    height: 40px;                       /*设置高度*/
    font-size: 18px;                    /*设置文字大小*/
    border-radius: 5px;                 /*设置圆角边框*/
    cursor: pointer;                    /*设置鼠标形状*/
    color: #fff;                        /*设置文字颜色*/
    border: none;                       /*设置无边框*/
}
```

```css
    .editing button.light {
        background-color: #0f88eb;                    /*设置背景颜色*/
        margin-right: 5px;                            /*设置右外边距*/
    }
    .editing button.dark {
        background-color: #ccc;                       /*设置背景颜色*/
        margin-right: 5px;                            /*设置右外边距*/
    }
</style>
```

7.9 路由配置

下面给出该项目的路由配置，包括引入页面组件、定义路由、创建路由对象、设置路由跳转后页面置顶，以及当路由发生变化时修改页面标题的功能。代码如下：

```js
import { createRouter, createWebHistory } from 'vue-router'
import HomeTop from "@/components/HomeTop.vue";
import UserLogin from "@/components/UserLogin.vue";
import UserRegister from "@/components/UserRegister.vue";
import ContactHome from "@/components/ContactHome.vue";
import ShowContacts from "@/components/ShowContacts.vue";
import AddCon from "@/components/AddCon.vue";
import SaveUser from "@/components/SaveUser.vue";
const routes = [
  {
    path: '/',
    redirect: '/home/register'
  },
  {
    path: '/home',
    component: HomeTop,
    children: [
      {
        path: 'login',
        component: UserLogin,
        meta: {
          title: '用户登录'
        }
      },
      {
        path: 'register',
        component: UserRegister,
        meta: {
          title: '用户注册'
        }
      }
    ]
  },
  {
    path: '/contacts',
    component: ContactHome,
    children: [
      {
        path: '',
        component: ShowContacts,
        meta: {
          logined: true,
          title: '通讯录'
```

```
          }
        },
        {
          path: 'add',
          component: AddCon,
          meta: {
            title: '添加联系人'
          }
        },
        {
          path: 'user',
          component: SaveUser,
          meta: {
            title: '个人中心'
          }
        },
      ]
    }
  ]
const router = createRouter({
  history: createWebHistory(process.env.BASE_URL),
  routes,
  //跳转页面后置项
  scrollBehavior(to,from,savedPosition){
    if(savedPosition){
      return savedPosition;
    }else{
      return {top:0,left:0}
    }
  }
})
router.beforeEach((to, from, next) => {
  /*路由发生变化时修改页面title */
  if (to.meta.title) {
    document.title = to.meta.title
  }
  next()
})
export default router
```

7.10 项目运行

通过前述步骤，我们已经设计并完成了"畅联通讯录"项目的开发。接下来，我们运行该项目，以检验我们的开发成果。首先打开命令提示符窗口，切换到项目所在目录，执行 npm run serve 命令运行项目，如图 7.22 所示。

在浏览器地址栏中输入 http://localhost:8080，然后按 Enter 键，进入用户注册页面。在该页面中，用户需要分别输入用户名、电话号码和密码，如图 7.23 所示。

单击"注册畅联"按钮，如果输入的内容符合要求，并且用户名未被使用，系统将会提示注册成功，如图 7.24 所示。

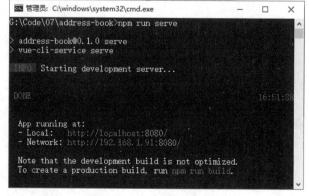

图 7.22 运行项目

图 7.23　用户注册页面

图 7.24　注册成功

单击"登录"超链接后，将进入用户登录页面。在登录页面输入正确的用户名和密码后，单击"登录"按钮，用户将被引导至通讯录页面。在该页面中，用户可以分页查看通讯录中的联系人信息，还可以对指定联系人进行编辑或删除的操作，如图 7.25 所示。

图 7.25　通讯录页面

7.11　源码下载

源码下载

本章虽然详细地讲解了如何编码实现"畅联通讯录"的各个功能，但给出的代码都是代码片段，而非完整的源代码。为了方便读者学习，本书提供了该项目的完整源代码，读者可以扫描右侧的二维码进行下载。

第 8 章
仿饿了么 APP

——Vue CLI + Router + axios + JSON Server + localStorage + SessionStorage

随着社会与经济的发展，人们生活和工作的节奏不断加快，网上订餐已经变得越来越普遍。外卖在大多数城市中发展十分迅速，以美团、饿了么为首的一些外卖平台已经家喻户晓。本章将使用 Vue CLI、axios、JSON Server 等技术开发一个移动端外卖应用平台——仿饿了么 APP。

本项目的核心功能及实现技术如下：

项目微视频

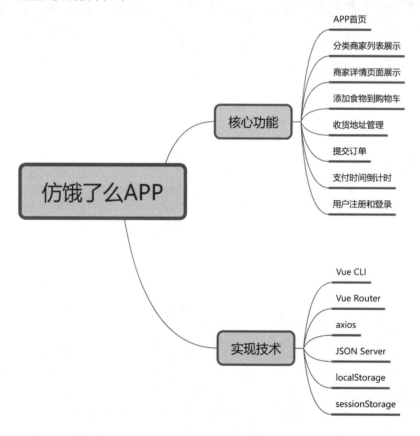

8.1 开发背景

随着移动互联网技术的快速发展，以及快递物流配送业务的日益完善，使得网络购物成为越来越受欢迎的新型消费方式。为了提高餐饮企业的业务水平和效率，实现比电话订餐更有效的订餐模式，网上订餐系统应运而生。目前，点外卖已经成为部分消费者的重要就餐方式，而饿了么、美团是最常见的两大外卖平台。本章将使用 Vue CLI、axios、JSON Server 等技术开发一个移动端外卖应用——仿饿了么 APP，其实现目标如下：
- ☑ 首页展示商家分类和商家信息列表。
- ☑ 商家详情页面展示商家信息和食物信息。
- ☑ 在商家详情页面，用户可以选择食物并将其添加到购物车中。
- ☑ 在收货地址列表中，用户可以选择默认收货地址。
- ☑ 用户提交订单后，系统生成用户订单。
- ☑ 支付页面将实现支付功能，同时给出支付时间倒计时提示。
- ☑ 提供用户注册和登录功能。

8.2 系 统 设 计

8.2.1 开发环境

本项目的开发及运行环境如下：
- ☑ 操作系统：推荐 Windows 10、Windows 11 或更高版本，同时兼容 Windows 7（SP1）。
- ☑ 开发工具：WebStorm。
- ☑ 开发框架：Vue.js 3.0。

8.2.2 业务流程

在设计仿饿了么 APP 时，用户如果未注册或已经注册但是未登录，则只能浏览首页，以及查看商家信息，而只有成功登录用户才能进行选择食物、更改购物车中的食物数量、选择收货地址和提交订单等操作。根据该项目的业务需求，我们设计如图 8.1 所示的业务流程图。

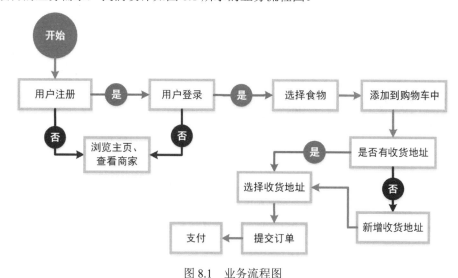

图 8.1 业务流程图

8.2.3 功能结构

本项目的功能结构已经在章首页中给出。其实现的具体功能如下：

- ☑ 首页：展示商家分类和商家信息列表，单击商家分类查看指定分类下的商家列表，单击商家信息进入商家详情页面。
- ☑ 分类商家列表页面：展示商家信息，包括商家图片、商家名称、起送费、配送费和商家描述等信息。
- ☑ 商家详情页面：展示商家信息和食物列表信息，用户可以通过单击"+"或"-"按钮来调整购物车中的食物数量。
- ☑ 确认订单页面：显示收货地址信息和购物车中的食物信息。
- ☑ 地址管理页面：用于新增收货地址、编辑收货地址或删除收货地址。
- ☑ 支付页面：显示支付时间倒计时、订单总价、商家名称、支付方式等信息，提供"确认支付"按钮。
- ☑ 订单列表页面：展示未支付和已支付的订单信息。
- ☑ 我的页面：展示当前登录用户的红包、卡券和地址等选项。
- ☑ 注册和登录页面：为用户提供注册和登录的服务功能。

8.3 技术准备

在开发仿饿了么 APP 时，我们主要应用 Vue CLI 脚手架工具、Vue Router、axios、JSON Server、localStorage 和 sessionStorage 等技术。Vue CLI 脚手架工具、Vue Router、axios、localStorage 和 sessionStorage 已经在前面的章节做了简要介绍。其中，Vue CLI 脚手架工具、Vue Router 和 axios 在《Vue.js 从入门到精通》中有详细的讲解，对这些知识不太熟悉的读者，可以参考该书对应的内容。下面将对 JSON Server 进行必要介绍，以确保读者可以顺利完成本项目。

1. JSON Server 简介

JSON Server 是一个用于快速搭建本地 RESTful API 的工具，它使用 JSON 文件作为数据源，并提供了一组简单的路由和端点，以此来模拟后端服务器的行为。作为一款小巧的接口模拟工具，JSON Server 能够帮助开发者快速搭建一套 RESTful 风格的 API，非常适合用于前端接口测试。在使用 JSON Server 时，只需指定一个 JSON 文件作为 API 的数据源，便可以轻松地开始使用它。

> **说明**
>
> RESTful API 是一种轻量级的 Web 服务，它可以通过 HTTP 请求进行访问，并使用 JSON 或 XML 等格式进行数据交换。RESTful API 通常使用标准的 HTTP 方法（GET、POST、PUT、DELETE 等）来实现对 Web 资源的各种操作，如获取数据、创建数据、更新数据和删除数据等。

2. 安装 JSON Server

要安装 JSON Server，需要先安装 Node.js。安装好 Node.js 后，就可以使用 npm 命令安装 JSON Server。在命令提示符窗口中输入以下命令：

```
npm install -g json-server
```

3. 启动 JSON Server

在指定目录下创建一个 db.json 文件作为数据源，在该文件中添加模拟服务器的数据。db.json 文件的代码如下：

```
{
  "users": [
    { "id": 1, "name": "Tony", "age": 25 },
```

```
    { "id": 2, "name": "Kelly", "age": 25 }
  ]
}
```

在命令提示符窗口中,将目录切换到 db.json 文件所在的目录,然后输入如下命令启动 JSON Server:

```
json-server --watch db.json
```

JSON Server 默认使用 3000 端口,如果想要修改端口号,可以在命令中添加--port 参数。例如,要将端口号改为 3030,可以使用以下命令:

```
json-server --watch db.json --port 3030
```

4. 使用 axios 实现数据请求

1)查询数据

要查询 JSON Server 中的数据,需要使用 GET 方式发送请求。例如,要查询 db.json 文件中所有用户的数据,可以使用以下代码:

```
axios.get('http://localhost:3000/users').then((response) => {
    console.log(response.data)
})
```

再如,要查询 db.json 文件中 id 值为 2 的数据,可以使用以下代码:

```
axios.get('http://localhost:3000/users?id=2').then((response) => {
    console.log(response.data)
})
```

再如,要查询 db.json 文件中 name 值是 Tony、age 值是 25 的数据,可以使用以下代码:

```
axios.get('http://localhost:3000/users?name=Tony&age=25').then((response) => {
    console.log(response.data)
})
```

2)添加数据

要向 JSON Server 中添加数据,需要使用 POST 方式发送请求。例如,要向 db.json 文件中添加一条数据,可以使用以下代码:

```
axios.post('http://localhost:3000/users',{
    name: 'Alice',
    age: 22
}).then((response) => {
    console.log(response.data)
})
```

3)更新数据

更新数据分为两种情况:一种是更新某个数据的所有字段,另一种是更新某个数据的部分字段。要更新所有字段,需要使用 PUT 方式发送请求;而仅更新部分字段,则应使用 PATCH 方式发送请求。例如,要将 db.json 文件中 id 为 1 的数据中的 name 值改为 Jack,age 值改为 30,可以使用以下代码:

```
axios.put('http://localhost:3000/users/1',{
    name: 'Jack',
    age: 30
}).then((response) => {
    console.log(response.data)
})
```

再如,要将 db.json 文件中 id 为 2 的数据中的 name 值改为 Rose,可以使用以下代码:

```
axios.patch('http://localhost:3000/users/2',{
```

```
    name: 'Rose'
}).then((response) => {
    console.log(response.data)
})
```

4）删除数据

要删除 JSON Server 中的数据，需要使用 DELETE 方式发送请求。例如，要删除 db.json 文件中 id 为 1 的数据，可以使用以下代码：

```
axios.delete('http://localhost:3000/users/1').then((response) => {
    console.log(response.data)
})
```

> **说明**
> 在该项目中，我们将所有需要用到的数据保存在 data.json 文件中，并将该文件放置在项目根目录中。在安装并启动 JSON 服务器后，我们使用 axios 发送请求到 data.json 文件以实现查询、添加、更新或删除数据的操作。

8.4 首页的设计与实现

仿饿了么 APP 的首页主要由商家分类、推荐商家列表和底部导航栏组成。首页的页面效果如图 8.2 所示。

8.4.1 商家分类页面设计

在仿饿了么 APP 首页中，商家分类一共有 10 个，其页面如图 8.3 所示。

图 8.2　仿饿了么 APP 首页效果

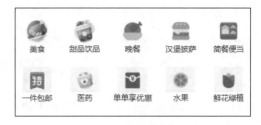

图 8.3　商家分类页面

具体实现步骤如下。

（1）在 views 文件夹下新建 IndexHome.vue 文件。在<template>标签中定义 ul 列表，在该列表中使用 v-for 指令对商家分类列表进行遍历，当用户单击某个商家分类时，将调用 toBusinessList()方法。关键代码如下：

```html
<!-- 商家分类 -->
<ul class="food-type">
  <li v-for="item in businessType" :key="item.typeId" @click="toBusinessList(item.typeId)">
    <img :src="item.typeImg">
    <p>{{ item.typeName }}</p>
  </li>
</ul>
```

（2）在<script>标签中定义商家分类列表，然后定义 toBusinessList()方法，调用该方法将会跳转到指定分类的商家列表页面。代码如下：

```js
export default {
    name: "IndexHome",
    data() {
      return {
        businessType: [                                       //商家分类信息
          {
            typeId: 1,
            typeImg: require('@/assets/type01.png'),
            typeName: '美食'
          },
          {
            typeId: 2,
            typeImg: require('@/assets/type02.png'),
            typeName: '甜品饮品'
          },
          {
            typeId: 3,
            typeImg: require('@/assets/type03.png'),
            typeName: '晚餐'
          },
          {
            typeId: 4,
            typeImg: require('@/assets/type04.png'),
            typeName: '汉堡披萨'
          },
          {
            typeId: 5,
            typeImg: require('@/assets/type05.png'),
            typeName: '简餐便当'
          },
          {
            typeId: 6,
            typeImg: require('@/assets/type06.png'),
            typeName: '一件包邮'
          },
          {
            typeId: 7,
            typeImg: require('@/assets/type07.png'),
            typeName: '医药'
          },
          {
            typeId: 8,
            typeImg: require('@/assets/type08.png'),
            typeName: '单单享优惠'
          },
```

```
                {
                    typeId: 9,
                    typeImg: require('@/assets/type09.png'),
                    typeName: '水果'
                },
                {
                    typeId: 10,
                    typeImg: require('@/assets/type10.png'),
                    typeName: '鲜花绿植'
                }
            ],
            business: {}                                     //商家信息
        };
    },
    methods: {
        toBusinessList(businessTypeId){                      //跳转到指定分类的商家列表页面
            this.$router.push({
                path: "/businessList",
                query: { businessTypeId: businessTypeId },
            });
        }
    }
}
```

8.4.2 推荐商家列表页面设计

在仿饿了么 APP 首页中，商家分类页面下方是推荐商家列表页面。每个商家都包括商家图片、商家名称、用户评分、起送费和配送费等信息，其页面如图 8.4 所示。

具体实现步骤如下。

（1）在 IndexHome.vue 文件的<template>标签中定义 ul 列表，在该列表中，使用 v-for 指令对推荐商家列表进行遍历。当用户单击某个商家时，将调用 toBusinessInfo()方法，并传递商家 id 作为参数。关键代码如下：

图 8.4 推荐商家列表页面

```
<!-- 推荐商家列表 -->
<ul class="business">
    <li v-for="item in business" :key="item.id" @click="toBusinessInfo(item.businessId)">
        <img :src="item.businessImg">
        <div class="business-info">
            <div class="business-info-head">
                <h3>{{ item.businessName }}</h3>
                <div class="business-info-like">
                    <div>&#8226;</div>
                    <div>&#8226;</div>
                    <div>&#8226;</div>
                </div>
            </div>
            <div class="business-info-star">
                <div class="business-info-star-left">
                    <p>{{ item.score }}分 月售{{ item.sales }}+</p>
                </div>
                <p>{{ item.time }} {{item.distance}}</p>
            </div>
            <div class="business-info-delivery">
                <p>起送&#165;{{ item.startPrice }} 
                    <span v-if="item.deliveryPrice>0">配送&#165;{{ item.deliveryPrice }}</span>
                    <span v-else>免配送费</span>
```

```html
        </p>
        <p class="business-info-delivery-right">{{ item.delivery }}</p>
      </div>
      <div class="business-info-promotion">
        <div class="business-info-promotion-left">
          <p>{{ item.businessExplain }}</p>
        </div>
      </div>
      <div class="business-info-promotion">
        <div class="business-info-promotion-left">
          <span v-for="t in item.preferential" :key="t">{{ t }}</span>
        </div>
        <div class="business-info-promotion-right">
          <i class="fa fa-caret-down"></i>
        </div>
      </div>
    </div>
  </li>
</ul>
```

（2）在<script>标签中引入 axios 并创建 axios 实例，在 mounted 选项中调用 init()方法，该方法使用 axios 实例发送 GET 请求到服务器，以获取所有商家列表，并将获取到的数据保存在 business 中。然后定义 toBusinessInfo()方法，该方法在被调用时，会根据传入的商家 ID 将页面跳转到商家详情页面。代码如下：

```javascript
import axios from "axios";                              //引入 axios
const api = axios.create({                              //创建 axios 实例
    baseURL: 'http://localhost:3000'
});
export default {
    name: "IndexHome",
    data() {
        return {
            business: {}                                //商家信息
        };
    },
    mounted() {
        this.init();                                    //页面载入后调用 init()方法查找商家信息
    },
    methods: {
        init(){
            api.get(`/BusinessList`).then((response) => {   //查找商家列表
                this.business = response.data;
            }).catch((error) => {
                console.error(error);
            });
        },
        toBusinessInfo(businessId) {                    //跳转到商家详情页面
            this.$router.push({
                path: "/businessInfo",
                query: { businessId: businessId },
            });
        }
    }
}
```

8.4.3 底部导航栏的设计

底部导航栏主要包括"首页""订单"和"我的"这 3 个选项，其页面效果如图 8.5 所示。

图 8.5 底部导航栏的页面效果

具体实现步骤如下。

（1）在 components 文件夹下新建一个名为 TheFooter.vue 的文件。在<template>标签中，定义一个 ul 列表，在该列表中使用<router-link>组件设置导航链接，并将<router-link>渲染为<li>标签。代码如下：

```html
<template>
    <ul class="footer">
        <router-link to="/index" custom v-slot="{navigate, isExactActive}">
            <li :class="[isExactActive && 'router-link-exact-active']" @click="navigate">
                <i class="fa fa-home"></i>
                <p>首页</p>
            </li>
        </router-link>
        <router-link to="/orderList" custom v-slot="{navigate, isExactActive}">
            <li :class="[isExactActive && 'router-link-exact-active']" @click="navigate">
                <i class="fa fa-file-text-o"></i>
                <p>订单</p>
            </li>
        </router-link>
        <router-link to="/my" custom v-slot="{navigate, isExactActive}">
            <li :class="[isExactActive && 'router-link-exact-active']" @click="navigate">
                <i class="fa fa-user-o"></i>
                <p>我的</p>
            </li>
        </router-link>
    </ul>
</template>
```

（2）在<style>标签中，编写底部导航栏组件的样式。代码如下：

```css
<style scoped>
.footer {
    width: 100%;                            /*设置宽度*/
    height: 14vw;                           /*设置高度*/
    border-top: solid 1px #ddd;             /*设置上边框*/
    background-color: #fff;                 /*设置背景颜色*/
    position: fixed;                        /*设置相对于浏览器窗口进行定位*/
    left: 0;                                /*设置元素距离左端距离*/
    bottom: 0;                              /*设置元素距离底端距离*/
    display: flex;                          /*设置弹性容器*/
    justify-content: space-around;          /*设置每个元素两侧的间隔相等*/
    align-items: center;                    /*设置每个元素在交叉轴上居中对齐*/
}
.wrapper .footer li {
    display: flex;                          /*设置弹性容器*/
    flex-direction: column;                 /*设置元素纵向排列*/
    justify-content: center;                /*设置元素在主轴上居中排列*/
    align-items: center;                    /*设置每个元素在交叉轴上居中对齐*/
    color: #222222;                         /*设置文字颜色*/
    user-select: none;                      /*设置不能选取文本*/
    cursor: pointer;                        /*设置鼠标形状*/
}
.wrapper .footer li p {
    font-size: 2.8vw;                       /*设置文字大小*/
}
.wrapper .footer li i {
    font-size: 5vw;                         /*设置文字大小*/
}
.router-link-exact-active i,.router-link-exact-active p{
    color: #FF0000;                         /*设置文字颜色*/
}
</style>
```

8.5 分类商家列表的设计与实现

在仿饿了么 APP 首页中，用户单击某个商家分类会跳转到对应的分类商家列表页面。分类商家列表页面展示了某个商家分类对应的商家信息，其中每个商家都包括商家图片、商家名称、起送费、配送费和商家描述等信息。其中，"美食"分类对应的商家列表如图 8.6 所示。

图 8.6 "美食"分类对应的商家列表

具体实现步骤如下。

（1）在 views 文件夹下新建 BusinessList.vue 文件。在<template>标签中定义 ul 列表，在该列表中使用 v-for 指令对分类商家列表进行遍历。当用户单击某个商家时，将调用 toBusinessInfo()方法，并传递商家 id 作为参数。代码如下：

```
<template>
  <div class="wrapper">
    <header>
      <div>
        <i class="fa fa-angle-left" @click="$router.go(-1)"></i>
      </div>
      <p>商家列表</p>
    </header>
    <ul class="business">
      <li v-for="item in businessArr" :key="item.businessId" @click="toBusinessInfo(item.businessId)">
        <div class="business-img">
          <img :src="item.businessImg" />
          <div class="business-img-quantity" v-show="item.quantity > 0">
            {{ item.quantity }}
```

```
          </div>
        </div>
        <div class="business-info">
          <h3>{{ item.businessName }}</h3>
          <p>
            &#165;{{ item.startPrice }}起送 |
            <span v-if="item.deliveryPrice>0">&#165;{{ item.deliveryPrice }}配送</span>
            <span v-else>免配送费</span>
          </p>
          <p>{{ item.businessExplain }}</p>
        </div>
      </li>
    </ul>
    <Footer></Footer>
  </div>
</template>
```

（2）在<script>标签中，引入 axios 并创建 axios 实例。在 data 选项中，定义商家分类 id、分类商家信息数组和当前登录用户信息。然后在 mounted 选项中使用 axios 发送 GET 请求，根据商家分类 id 查询商家信息，并将获取的商家列表保存在 businessArr 中。接着判断用户是否已登录，如果用户已登录，就调用 listCart()方法，该方法会根据用户手机号码查找购物车列表，统计该商家在购物车中的食物数量。最后定义toBusinessInfo()方法，调用该方法，将会跳转到商家详情页面。代码如下：

```
<script>
  import axios from "axios";                                              //引入 axios
  const api = axios.create({                                              //创建 axios 实例
    baseURL: 'http://localhost:3000'
  });
  import Footer from "../components/TheFooter.vue";
  export default {
    name: "BusinessList",
    data() {
      return {
        businessTypeId: this.$route.query.businessTypeId,                 //商家分类 id
        businessArr: [],                                                  //商家信息数组
        user: JSON.parse(sessionStorage.getItem("user"))                  //登录用户信息
      };
    },
    mounted() {
      //根据商家分类 id 查询商家信息
      api.get(`/BusinessList?businessTypeId=${this.businessTypeId}`).then((response) => {
        this.businessArr = response.data;
        //如果用户已登录，就查询购物车中是否已经选择了该商家的某种食物
        if (this.user) {
          this.listCart();
        }
      }).catch((error) => {
        console.error(error);
      });
    },
    components: {
      Footer,
    },
    methods: {
      listCart() {
        //根据用户手机号码查找购物车
        api.get(`/CartList?userId=${this.user.userId}`).then((response) => {
          let cartArr = response.data;
          //遍历商家列表
          for (let businessItem of this.businessArr) {
            businessItem.quantity = 0;
```

```
            //遍历购物车,统计该商家购物车中的食物数量
            for (let cartItem of cartArr) {
              if (cartItem.businessId == businessItem.businessId) {
                businessItem.quantity += cartItem.quantity;
              }
            }
          }
        }).catch((error) => {
          console.error(error);
        });
      },
      toBusinessInfo(businessId) {               //跳转到商家详情页面
        this.$router.push({
          path: "/businessInfo",
          query: { businessId: businessId },
        });
      }
    },
  };
</script>
```

8.6 商家详情页面的设计与实现

图 8.7 商家详情页面

在仿饿了么 APP 首页或分类商家列表页面中,单击某个商家名称会跳转到对应的商家详情页面。商家详情页面由两部分组成:商家信息界面和购物车界面。商家信息界面展示了该商家的具体信息,包括商家图片、商家名称、起送费、配送费、商家描述和该商家提供的食物等信息。用户在商家信息页面选择食物后,购物车页面将显示所选食物总数、总价、配送费等信息。商家详情页面的效果如图 8.7 所示。

8.6.1 商家信息页面设计

在商家详情页面中,商家信息主要包括商家图片、商家名称、起送费、配送费、商家描述和该商家提供的食物等信息。具体实现步骤如下。

(1)在 views 文件夹下新建 BusinessInfo.vue 文件。在 <template> 标签中定义 <div> 元素,在该元素中定义商家图片、商家名称、起送费、配送费和商家描述信息。然后定义一个 ul 列表,在该列表中使用 v-for 指令对商家提供的食物列表进行遍历。当用户单击食物右侧的"-"按钮时,将调用 minus() 方法;当用户单击食物右侧的"+"按钮时,将调用 add() 方法。关键代码如下:

```
<header>
  <div>
    <i class="fa fa-angle-left" @click="$router.go(-1)"></i>
  </div>
  <p>商家信息</p>
</header>
<div class="business-logo">
  <img :src="business.businessImg" />
```

```html
</div>
<div class="business-info">
  <h1>{{ business.businessName }}</h1>
  <p>
    &#165;{{ business.startPrice }}起送
    <span v-if="business.deliveryPrice > 0">&#165;{{business.deliveryPrice }}配送</span>
    <span v-else>免配送费</span>
  </p>
  <p>{{ business.businessExplain }}</p>
</div>
<ul class="food">
  <li v-for="(item, index) in foodArr" :key="item.foodId">
    <div class="food-left">
      <img :src="item.foodImg" />
      <div class="food-left-info">
        <h3>{{ item.foodName }}</h3>
        <p>{{ item.foodExplain }}</p>
        <p>&#165;{{ item.foodPrice }}</p>
      </div>
    </div>
    <div class="food-right">`
      <dlv>
        <i class="fa fa-minus-circle" @click="minus(index)" v-show="item.quantity !== 0"></i>
      </dlv>
      <p>
        <span v-show="item.quantity !== 0">{{ item.quantity }}</span>
      </p>
      <div>
        <i class="fa fa-plus-circle" @click="add(index)"></i>
      </div>
    </div>
  </li>
</ul>
```

（2）在<script>标签中引入 axios 并创建 axios 实例，在 data 选项中定义商家 id、商家信息、当前商家提供的食物信息数组和当前登录用户信息。然后在 mounted 选项中使用 axios 发送 GET 请求，根据商家 id 查询商家信息，并将获取的商家信息保存在 business 中，同时将获取的食物信息数组保存在 foodArr 中。接着判断用户是否已登录，如果用户已登录，就调用 listCart()方法，该方法会根据商家 id 和用户手机号码查找购物车列表，并获取购物车中指定食物的数量。然后定义用于更新食物数量的多个方法。代码如下：

```js
<script>
  import axios from "axios";                                              //引入 axios
  const api = axios.create({                                              //创建 axios 实例
    baseURL: 'http://localhost:3000'
  });
  export default {
    name: "BusinessInfo",
    data() {
      return {
        businessId: this.$route.query.businessId,                         //商家 id
        business: {},                                                     //商家信息
        foodArr: [],                                                      //食物信息数组
        user: JSON.parse(sessionStorage.getItem("user"))                  //登录用户信息
      };
    },
    mounted() {
      //根据商家 id 查找商家信息
      api.get(`/BusinessList?businessId=${this.businessId}`).then((response) => {
        this.business = response.data[0];
        this.foodArr = response.data[0].foodList;
        for (let foodItem of this.foodArr) {
```

```js
          foodItem.quantity = 0;
        }
        //如果用户已登录，就查询购物车中是否已经选择了某种食物
        if (this.user) {
          this.listCart();
        }
      }).catch((error) => {
        console.error(error);
      });
    },
    methods: {
      listCart() {
        //根据商家 id 和用户手机号码查找购物车
        api.get(`/CartList?businessId=${this.businessId}&userId=${this.user.userId}`).then((response) => {
          let cartArr = response.data;
          //遍历食物信息数组
          for (let foodItem of this.foodArr) {
            foodItem.quantity = 0;
            for (let cartItem of cartArr) {
              if (cartItem.foodId === foodItem.foodId) {
                foodItem.quantity = cartItem.quantity;
              }
            }
          }
        })
        .catch((error) => {
          console.error(error);
        });
      },
      add(index) {
        if (!this.user) {                                          //如果用户未登录，就跳转到登录页面
          this.$router.push({ path: "/login" });
          return;
        }
        if (this.foodArr[index].quantity === 0) {
          this.savaCart(index);                                    //将该食物添加到购物车中
        } else {
          this.updateCart(index, 1);                               //更新购物车中该食物数量，使其加 1
        }
      },
      minus(index) {
        if (!this.user) {                                          //如果用户未登录，就跳转到登录页面
          this.$router.push({ path: "/login" });
          return;
        }
        if (this.foodArr[index].quantity > 1) {
          this.updateCart(index, -1);                              //更新购物车中该食物的数量，使其减 1
        } else {
          this.removeCart(index);                                  //从购物车中移除该食物
        }
      },
      savaCart(index) {
        api.post("/CartList", {                                    //向购物车中添加食物
          businessId: this.businessId,
          userId: this.user.userId,
          foodId: this.foodArr[index].foodId,
          quantity: 1,
          foodInfo: {
            foodImg: this.foodArr[index].foodImg,
            foodName: this.foodArr[index].foodName,
            foodPrice: this.foodArr[index].foodPrice
          }
```

```js
            }).then((response) => {
                if (response.data.id) {
                    this.foodArr[index].quantity = 1;                    //将食物数量设置为 1
                } else {
                    alert("向购物车中添加食物失败！");
                }
            }).catch((error) => {
                console.error(error);
            });
        },
        updateCart(index, num) {
            //根据指定商家 id 和指定食物 id 查找购物车，以获取购物车 id
            api.get(`/CartList?userId=${this.user.userId}&businessId=${this.businessId}
                &foodId=${this.foodArr[index].foodId}`).then((response) => {
                api.patch(`/CartList/${response.data[0].id}`, {           //根据购物车 id 更新食物数量
                    "quantity": this.foodArr[index].quantity + num,
                }).then((response) => {
                    if (response.data.id) {
                        //将该食物数量加 1 或减 1
                        this.foodArr[index].quantity += num;
                    } else {
                        alert("向购物车中更新食物失败！");
                    }
                }).catch((error) => {
                    console.error(error);
                });
            }).catch((error) => {
                console.error(error);
            });
        },
        removeCart(index) {
            //根据指定 id 的食物查找购物车，以获取购物车 id
            api.get(`/CartList?userId=${this.user.userId}&businessId=${this.businessId}
                &foodId=${this.foodArr[index].foodId}`).then((response) => {
                //根据购物车 id 从购物车中删除该食物
                api.delete(`/CartList/${response.data[0].id}`).then((response) => {
                    if (response.data.id) {
                        //将该食物数量设置为 0
                        this.foodArr[index].quantity = 0;
                    } else {
                        alert("从购物车中删除食物失败！");
                    }
                }).catch((error) => {
                    console.error(error);
                });
            })
        }
    }
};
</script>
```

8.6.2 购物车页面设计

在商家详情页面的下方是购物车页面，其中展示了用户在该商家提供的食物中选中的食物总数量、食物总价、配送费，同时还提供了一个"去结算"按钮。具体实现步骤如下。

（1）在 BusinessInfo.vue 文件的<template>标签中定义<div>元素，在该元素中定义用户在该商家提供的食物中选中的食物总数量、食物总价、配送费和"去结算"按钮。当用户单击"去结算"按钮时，将调用 toOrder()方法。代码如下：

```
<!-- 购物车 -->
<div class="cart">
  <div class="cart-left">
    <div class="cart-left-icon":style="totalQuantity === 0 ? 'background-color:#505051;' : 'background-color:#3190E8;'" >
      <i class="fa fa-shopping-cart"></i>
      <div class="cart-left-icon-quantity" v-show="totalQuantity !== 0">
        {{ totalQuantity }}
      </div>
    </div>
    <div class="cart-left-info">
      <p>&#165;{{ totalPrice }}</p>
      <p v-if="business.deliveryPrice>0">另需配送费{{ business.deliveryPrice }}元</p>
      <p v-else>免配送费</p>
    </div>
  </div>
  <div class="cart-right">
    <!-- 未达到起送费 -->
    <div class="cart-right-item" v-show="totalSettle < business.startPrice"
        style="background-color: #535356; cursor: default">
      &#165;{{ business.startPrice }}起送
    </div>
    <!-- 达到起送费 -->
    <div class="cart-right-item" @click="toOrder" v-show="totalSettle >= business.startPrice">
      去结算
    </div>
  </div>
</div>
```

（2）在<script>标签中定义 toOrder()方法，调用该方法时，会跳转到确认订单页面。然后定义 totalPrice、totalQuantity 和 totalSettle 计算属性，分别获取食物总价格、食物总数量和购物车的总价格。代码如下：

```
export default {
    methods: {
        toOrder() {                                              //跳转到确认订单页面
            this.$router.push({
                path: "/orders",
                query: { businessId: this.business.businessId },
            });
        },
    },
    computed: {
        totalPrice() {                                           //食物总价格
            let total = 0;
            for (let item of this.foodArr) {
                total += item.foodPrice * item.quantity;
            }
            return total;
        },
        totalQuantity() {                                        //食物总数量
            let quantity = 0;
            for (let item of this.foodArr) {
                quantity += item.quantity;
            }
            return quantity;
        },
        totalSettle() {                                          //总价格 = 食物总价格 + 配送费
            return this.totalPrice + this.business.deliveryPrice;
        },
    }
};
```

8.7 确认订单页面的设计与实现

在商家详情页面中,当用户单击"去结算"按钮时,系统将引导用户跳转到确认订单页面。确认订单页面展示了订单的具体信息,包括收货人信息、商家名称、购买的食物信息、配送费、订单总价格和"提交订单"按钮,其页面效果如图 8.8 所示。

8.7.1 确认订单页面设计

确认订单页面的具体实现步骤如下。

(1)在 views 文件夹下新建 UserOrders.vue 文件。在 <template>标签中定义<div>元素,在该元素中分别定义收货人信息、购物车信息、订单总价格和"提交订单"按钮。当用户单击收货人信息界面时,将调用 toUserAddress()方法;当用户单击"提交订单"按钮时,将调用 toPayment()方法。代码如下:

图 8.8 确认订单页面

```
<template>
  <div class="wrapper">
    <header>
      <div>
        <i class="fa fa-angle-left" @click="$router.go(-1)"></i>
      </div>
      <p>确认订单</p>
    </header>
    <div class="order-info">
      <div class="order-info-address" @click="toUserAddress">
        <p>
          {{
            deliveryAddress !== null ? deliveryAddress.address : "请选择收货地址"
          }}
        </p>
        <i class="fa fa-angle-right"></i>
      </div>
      <p v-if="deliveryAddress !== null">
        {{ deliveryAddress.contactName }}({{ deliveryAddress.contactSex === 1 ? "先生" : "女士" }})
        {{ deliveryAddress.contactTel }}
      </p>
    </div>
    <div class="order-business-info">
      <h3>{{ business.businessName }}</h3>
      <ul class="order-detailed">
        <li v-for="item in cartArr" :key="item.id">
          <div class="order-detailed-left">
            <img :src="item.foodInfo.foodImg">
            <p>{{ item.foodInfo.foodName }} x {{ item.quantity }}</p>
          </div>
          <p>&#165;{{ item.foodInfo.foodPrice * item.quantity }}</p>
        </li>
      </ul>
      <div class="order-deliveryfee">
```

```html
        <p>配送费</p>
        <p>&#165;{{ business.deliveryPrice }}</p>
      </div>
    </div>
    <div class="total">
      <div class="total-left">合计 &#165;{{ totalPrice }}</div>
      <div class="total-right" @click="toPayment">提交订单</div>
    </div>
  </div>
</template>
```

（2）在<script>标签中引入 axios 并创建 axios 实例，在 data 选项中定义当前登录用户信息、默认收货地址信息、商家 id、商家信息和指定的购物车列表信息。在 mounted 选项中使用 axios 发送 GET 请求，根据商家 id 查询商家信息，并将获取的商家信息保存在 business 中。接着根据用户手机号码和商家 id 查询用户在该商家中的购物车列表，并将获取的购物车列表保存在 cartArr 中。之后定义 toUserAddress()方法和 toPayment()方法。调用 toUserAddress()方法会跳转到当前登录用户收货地址管理页面；调用 toPayment()方法会创建用户订单，将订单信息添加到订单列表中，并跳转到支付页面。代码如下：

```javascript
<script>
  import axios from "axios";                                        //引入 axios
  const api = axios.create({                                        //创建 axios 实例
    baseURL: 'http://localhost:3000'
  });
  export default {
    name: "UserOrders",
    data() {
      return {
        user: JSON.parse(sessionStorage.getItem("user")),           //登录用户信息
        deliveryAddress: JSON.parse(localStorage.getItem("defaultAddress")),  //默认收货地址
        businessId: this.$route.query.businessId,                   //商家 id
        business: {},                                               //商家信息
        cartArr: []                                                 //购物车列表
      };
    },
    mounted() {
      //根据商家 id 查询当前商家信息
      api.get(`/BusinessList?businessId=${this.businessId}`).then((response) => {
        this.business = response.data[0];
      }).catch((error) => {
        console.error(error);
      });
      //根据用户手机号码和商家 id 查询用户在购物车中的当前商家食物列表
      api.get(`/CartList?userId=${this.user.userId}&businessId=${this.businessId}`).then((response) => {
        this.cartArr = response.data;
      }).catch((error) => {
        console.error(error);
      });
    },
    methods: {
      toUserAddress(){
        this.$router.push({                                         //跳转到用户收货地址页面
          path: "/userAddress",
          query: { businessId: this.businessId },
        });
      },
      toPayment(){
        if (!this.deliveryAddress) {
          alert("请选择送货地址！ ");
          return;
```

```
            api.post("/OrderList", {                           //创建订单，将订单信息添加到订单列表中
              userId: this.user.userId,                        //用户手机号码
              business: {                                      //商家信息
                businessId: this.businessId,
                businessName: this.business.businessName,
                deliveryPrice: this.business.deliveryPrice
              },
              address: this.deliveryAddress,                   //收货地址信息
              food: this.cartArr,                              //购物车信息
              orderTotal: this.totalPrice,                     //总价
              totalNum: this.totalNum,                         //订单中的食物数量
              orderState: 0                                    //订单状态
            }).then((response) => {
              if (response.data.id) {
                //跳转到支付页面
                this.$router.push({ path: "/payment", query: { orderId: response.data.id } });
              } else {
                alert("创建订单失败！");
              }
            }).catch((error) => {
              console.error(error);
            });
          }
        },
        computed: {
          totalPrice(){                                        //订单总价
            let totalPrice = 0;
            for (let cartItem of this.cartArr) {
              totalPrice += cartItem.foodInfo.foodPrice * cartItem.quantity;
            }
            totalPrice += this.business.deliveryPrice;
            return totalPrice;
          },
          totalNum(){                                          //订单中的食物数量
            let totalNum = 0;
            for (let cartItem of this.cartArr) {
              totalNum += cartItem.quantity;
            }
            return totalNum;
          }
        }
      }
</script>
```

8.7.2 新增收货地址页面的设计

如果当前登录用户尚未设置默认的收货地址，那么确认订单页面的上方会显示"请选择收货地址"。当用户单击"请选择收货地址"后，用户将被引导至地址管理页面。如果用户尚未添加任何收货地址，该页面会显示"新增收货地址"按钮。单击该按钮后，用户将被引导至新增收货地址页面，在该页面中，用户可以填写联系人姓名、选择联系人性别、填写联系人电话，以及填写具体的收货地址。新增收货地址页面的效果如图 8.9 所示。

具体实现步骤如下。

（1）在 views 文件夹下新建 AddUserAddress.vue 文件。在<template>标签中定义 ul 列表，在该列表中添加表单元素，

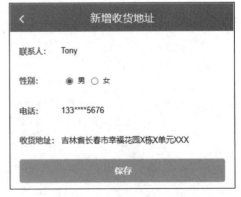

图 8.9　新增收货地址页面

用于输入联系人姓名、选择联系人性别、输入联系人电话，以及输入具体的收货地址。然后添加一个"保存"按钮，当用户单击该按钮时，将调用 addUserAddress()方法。代码如下：

```
<template>
  <div class="wrapper">
    <header>
      <div>
        <i class="fa fa-angle-left" @click="$router.go(-1)"></i>
      </div>
      <p>新增收货地址</p>
    </header>
    <ul class="form-box">
      <li>
        <div class="title">联系人：</div>
        <div class="content">
          <input type="text" v-model="deliveryAddress.contactName" placeholder="联系人姓名">
        </div>
      </li>
      <li>
        <div class="title">性别：</div>
        <div class="content" style="font-size: 3vw">
          <input type="radio" v-model="deliveryAddress.contactSex" value="1" style="width: 6vw; height: 3.2vw">男
          <input type="radio" v-model="deliveryAddress.contactSex" value="0" style="width: 6vw; height: 3.2vw">女
        </div>
      </li>
      <li>
        <div class="title">电话：</div>
        <div class="content">
          <input type="tel" v-model="deliveryAddress.contactTel" placeholder="电话">
        </div>
      </li>
      <li>
        <div class="title">收货地址：</div>
        <div class="content">
          <input type="text" v-model="deliveryAddress.address" placeholder="收货地址">
        </div>
      </li>
    </ul>
    <div class="button-add">
      <button @click="addUserAddress">保存</button>
    </div>
    <Footer></Footer>
  </div>
</template>
```

（2）在<script>标签中引入 axios 并创建 axios 实例。接着，在 data 选项中定义当前登录用户信息、商家 id 和收货地址信息。然后定义 addUserAddress()方法，该方法将验证用户输入的内容是否为空。如果输入的内容完整，就使用 axios 发送 POST 请求，将收货人信息添加到收货地址列表中，并自动跳转到确认订单页面。代码如下：

```
<script>
  import Footer from "../components/TheFooter.vue";           //引入底部组件
  import axios from "axios";                                  //引入 axios
  const api = axios.create({                                  //创建 axios 实例
    baseURL: 'http://localhost:3000'
  });
  export default {
    name: "BusinessInfo",
    data() {
      return {
        user: JSON.parse(sessionStorage.getItem("user")),     //登录用户信息
```

```
              businessId: this.$route.query.businessId,        //商家id
              deliveryAddress:{                                //收货地址信息
                contactName: "",
                contactSex: 1,
                contactTel: "",
                address: ""
              }
            }
        },
        components: {
          Footer
        },
        methods: {
          addUserAddress(){                                    //增加收货地址
            if (this.deliveryAddress.contactName === "") {
              alert("联系人姓名不能为空！");
              return;
            }
            if (this.deliveryAddress.contactTel === "") {
              alert("联系人电话不能为空！");
              return;
            }
            if (this.deliveryAddress.address === "") {
              alert("联系人地址不能为空！");
              return;
            }
            this.deliveryAddress.userId = this.user.userId;
            api.post("/DeliveryAddressList", this.deliveryAddress).then((response) => {
              if (response.data.id) {
                localStorage.removeItem("defaultAddress");     //删除默认地址
                this.$router.push({                            //跳转到确认订单页面
                  path: "/orders",
                  query: { businessId: this.businessId },
                });
              } else {
                alert("新增收获地址失败！");
              }
            }).catch((error) => {
              console.error(error);
            });
          }
        }
      }
</script>
```

8.7.3 地址管理页面的设计

在新增收货地址页面，单击"保存"按钮后，用户将被重定向至地址管理页面。该页面显示了当前用户添加的收货人信息，并允许用户对某个收货人信息进行编辑或删除。该页面效果如图 8.10 所示。

具体实现步骤如下。

（1）在 views 文件夹下新建 UserAddress.vue 文件。在 <template>标签中定义 ul 列表，在该列表中对当前用户添加的收货人信息列表进行遍历。当用户单击收货人信息时，将调用 setDefaultDeliveryAddress()方法；当用户单击右侧的编辑图标和删除图标时，将分别调用 editUserAddress()

图 8.10 地址管理页面

方法和 removeUserAddress()方法。然后添加一个"新增收货地址"按钮，当用户单击该按钮时，将调用 toAddUserAddress()方法。代码如下：

```html
<template>
  <div class="wrapper">
    <header>
      <div>
        <i class="fa fa-angle-left" @click="$router.go(-1)"></i>
      </div>
      <p>地址管理</p>
    </header>
    <ul class="address-list">
      <li v-for="item in deliveryAddressArr" :key="item.id">
        <div class="address-list-left" @click="setDefaultDeliveryAddress(item)">
          <p>{{ item.address }}</p>
          <h5>
            {{ item.contactName }}（{{ item.contactSex == 1 ? "先生" : "女士" }}）
            {{ item.contactTel }}
          </h5>
        </div>
        <div class="address-list-right">
          <i class="fa fa-edit" @click="editUserAddress(item.id)"></i>
          <i class="fa fa-remove" @click="removeUserAddress(item.id)"></i>
        </div>
      </li>
    </ul>
    <div class="add-btn" @click="toAddUserAddress">
      <i class="fa fa-plus-circle"></i>
      <p>新增收货地址</p>
    </div>
    <Footer></Footer>
  </div>
</template>
```

（2）在<script>标签中引入 axios 并创建 axios 实例，在 data 选项中分别定义当前登录用户信息、商家 id 和收货地址列表。然后在 mounted 选项中调用 DeliveryAddressListByUserId()方法，该方法会使用 axios 发送 GET 请求，根据当前登录用户的手机号码查询收货地址信息。接着定义 setDefaultDeliveryAddress()方法，该方法被调用时，它会将用户选择的地址作为默认收货地址保存在 localStorage 中，然后页面会跳转到提交订单页面。最后分别定义添加、编辑、删除收货地址的方法。代码如下：

```js
<script>
import Footer from "../components/TheFooter.vue";
import axios from "axios";                                      //引入 axios
const api = axios.create({                                      //创建 axios 实例
  baseURL: 'http://localhost:3000'
});
export default {
  name: "BusinessInfo",
  data() {
    return {
      user: JSON.parse(sessionStorage.getItem("user")),         //登录用户信息
      businessId: this.$route.query.businessId,                 //商家 id
      deliveryAddressArr: []                                    //收货地址列表
    };
  },
  components: {
    Footer
  },
  mounted() {
```

```js
      this.DeliveryAddressListByUserId();
    },
    methods: {
      DeliveryAddressListByUserId(){
        //根据用户 id 查询收货地址信息
        api.get(`/DeliveryAddressList?userId=${this.user.userId}`).then((response) => {
          this.deliveryAddressArr = response.data;
        }).catch((error) => {
          console.error(error);
        });
      },
      setDefaultDeliveryAddress(deliveryAddress) {
        //把用户选择的默认收货地址保存在 localStorage 中
        localStorage.setItem("defaultAddress", JSON.stringify(deliveryAddress));
        this.$router.push({                                           //跳转到提交订单页面
          path: "/orders",
          query: { businessId: this.businessId }
        });
      },
      toAddUserAddress(){
        this.$router.push({                                           //跳转到添加收货地址页面
          path: "/addUserAddress",
          query: { businessId: this.businessId }
        });
      },
      editUserAddress(id){
        this.$router.push({                                           //跳转到编辑收货地址页面
          path: "/editUserAddress",
          query: { businessId: this.businessId, id: id }
        });
      },
      removeUserAddress(id){
        if (!confirm("确认要删除此送货地址吗？")) {
          return;
        }
        //根据收货地址 id 删除收货地址
        api.delete(`/DeliveryAddressList/${id}`).then((response) => {
          if (response.data.id) {
            let defaultAddress = JSON.parse(localStorage.getItem("defaultAddress"));
            if (defaultAddress !== null && defaultAddress.id === id) {
              localStorage.removeItem("defaultAddress");              //移除默认收货地址
            }
            this.DeliveryAddressListByUserId();                       //重新获取收货地址列表
          } else {
            alert("删除地址失败！");
          }
        }).catch((error) => {
          console.error(error);
        });
      }
    }
  }
</script>
```

8.8 支付页面的设计与实现

在确认订单页面中，用户单击"提交订单"按钮，该页面会跳转到支付页面。该页面主要显示了支付时间倒计时、订单总价格、商家名称、支付方式和"确认支付"按钮，其页面效果如图 8.11 所示。

具体实现步骤如下。

（1）在 views 文件夹下新建 UserPayment.vue 文件。在<template>标签中定义<div>元素、<p>元素和 ul 列表，分别用于显示剩余支付时间、订单总价格和商家名称，以及支付方式。然后，添加一个"确认支付"按钮，当用户单击该按钮时，将调用 confirmPay()方法，并传递订单 id 作为参数。代码如下：

图 8.11 支付页面

```
<template>
  <div class="wrapper">
    <header>
      <p>在线支付</p>
    </header>
    <div class="remain-time">剩余支付时间：{{countdown}}</div>
    <p class="price">&#165;{{ orders.orderTotal }}</p>
    <p class="business-name">
      {{ businessInfo.businessName }}外卖订单
    </p>
    <ul class="payment-type">
      <li>
        <img src="../assets/alipay.png" />
        <i class="fa fa-check-circle"></i>
      </li>
      <li>
        <img src="../assets/wechat.png" />
      </li>
    </ul>
    <div class="payment-button">
      <button @click="confirmPay(orders.id)">确认支付</button>
    </div>
    <Footer></Footer>
  </div>
</template>
```

（2）在<script>标签中引入 axios 并创建 axios 实例，在 data 选项中定义数据，然后在 mounted 选项中获取支付剩余秒数，再使用 axios 发送 GET 请求，根据订单 id 查找订单信息，将获取的订单信息保存在 orders 中，将订单中的商家信息保存在 businessInfo 中。接着分别调用 startCountdown()方法和 showCountdown()方法，前者用于开始支付时间倒计时，后者显示支付时间倒计时。然后定义 confirmPay()方法，该方法会判断支付时间剩余秒数是否大于 0。如果大于 0，就根据订单 id 更新订单状态；否则就弹出"15 分钟内未支付，订单已自动取消"的提示，再根据订单 id 删除订单，并跳转到 APP 首页。代码如下：

```
<script>
  import Footer from "../components/TheFooter.vue";
  import axios from "axios";                                  //引入 axios
  const api = axios.create({                                  //创建 axios 实例
    baseURL: 'http://localhost:3000'
  });
  export default {
    name: "UserPayment",
    data() {
      return {
        orderId: this.$route.query.orderId,                   //订单 id
        orders: {},                                           //订单信息
        businessInfo: {},                                     //商家信息
        countdown: '',                                        //倒计时
        remainSeconds: 0,                                     //剩余秒数
```

```
      timer: null,                                                                    //超时 id
    }
  },
  components: {
    Footer
  },
  mounted() {
    //获取剩余秒数
    this.remainSeconds = localStorage.getItem(this.orderId) ? localStorage.getItem(this.orderId) : 15 * 60;
    //根据订单 id 查找订单信息
    api.get(`/OrderList?id=${this.orderId}`).then((response) => {
      this.orders = response.data[0];
      this.businessInfo = this.orders.business;                                        //获取订单中的商家信息
    }).catch((error) => {
      console.error(error);
    });
    //将当前 url 添加到 history 对象中
    history.pushState(null, null, document.URL);
    window.onpopstate = () => {                                                        //监听 history 对象的变化
      this.$router.push({ path: "/index" });
    };
    this.startCountdown();                                                             //开始倒计时
    this.showCountdown();                                                              //显示倒计时
  },
  unmounted() {
    window.onpopstate = null;
  },
  methods: {
    confirmPay(id){
      if(this.remainSeconds > 0){
        //根据订单 id 更新订单状态
        api.patch(`/OrderList/${id}`,{
          orderState: 1
        }).then((response) => {
          if(response.data.id){
            this.$router.push({ path: "/index" });                                     //跳转到主页
          }
        }).catch((error) => {
          console.error(error);
        });
      } else {
        alert("15 分钟内未支付，订单已自动取消");
        localStorage.removeItem(id);                                                   //删除 localStorage 中保存的剩余秒数
        //根据订单 id 删除订单
        api.delete(`/OrderList/${id}`).then((response) => {
          if(response.data.id){
            this.$router.push({ path: "/index" });                                     //跳转到首页
          }
        }).catch((error) => {
          console.error(error);
        });
      }
    },
    showCountdown(){
      let curMinutes = Math.floor(this.remainSeconds / 60);                            //获取倒计时中的分钟数
      let curSeconds = Math.floor(this.remainSeconds % 60 );                           //获取倒计时中的秒数
      curMinutes = curMinutes < 10 ? ('0' + curMinutes) : curMinutes;
      curSeconds = curSeconds < 10 ? ('0' + curSeconds) : curSeconds;
      this.countdown = "00:" + curMinutes + ":" + curSeconds;
    },
    startCountdown() {
      let t = this;
      this.timer = setInterval(() => {
```

```
          localStorage.setItem(this.orderId, t.remainSeconds);    //将剩余秒数保存在localStorage中
          if (t.remainSeconds > 0) {
            this.remainSeconds -= 1;                              //剩余秒数减1
            t.showCountdown();                                    //显示倒计时
          } else {
            t.clearCountdown();                                   //取消倒计时
          }
        }, 1000);
      },
      clearCountdown() {
        this.countdown = "00:00:00";
        clearInterval(this.timer);
        this.timer = null;
      }
    }
  }
</script>
```

> **说明**
> 在设计支付页面时,将支付剩余时间保存在 localStorage 中,并将订单 id 作为 key,这样可以实现多个订单页面显示支付时间倒计时的功能。

8.9 订单列表页面的设计与实现

在底部导航栏中,单击"订单"选项会跳转到订单列表页面。该页面展示了未支付和已支付的订单信息,包括商家名称、食物图片、支付状态和订单总价格等信息,其页面效果如图 8.12 所示。

具体实现步骤如下。

(1)在 views 文件夹下新建 OrderList.vue 文件。在 <template> 标签中定义 ul 列表。在 ul 列表中首先对未支付的订单列表进行遍历,在遍历时为每个订单添加一个"去支付"按钮,单击该按钮会调用 toPay() 方法,传递的参数是订单 id。然后对已支付的订单列表进行遍历,在遍历时为每个订单添加一个"再来一单"按钮,单击该按钮会调用 toBusinessInfo() 方法,传递的参数是订单中的商家 id。代码如下:

图 8.12 订单列表页面

```
<template>
  <div class="wrapper">
    <header>
      <div>
        <i class="fa fa-angle-left" @click="$router.go(-1)"></i>
      </div>
      <p>我的订单</p>
    </header>
    <ul class="order">
      <li v-for="order in UnpaidOrderArr" :key="order.id">
        <div class="order-info">
          <p @click="toBusinessInfo(order.business.businessId)">
            {{ order.business.businessName }}
            <i class="fa fa-caret-right"></i>
```

```
              </p>
              <p>未支付</p>
            </div>
            <div class="order-detail">
              <div>
                <span v-for="item in order.food" :key="item.id">
                  <img :src="item.foodInfo.foodImg">
                </span>
              </div>
              <div>
                <p>&#165;{{ order.orderTotal }}</p>
                <p>共{{order.totalNum}}件</p>
              </div>
            </div>
            <div class="order-info-right">
              <div class="order-info-right-icon" @click="toPay(order.id)">去支付</div>
            </div>
          </li>
          <li v-for="order in PaidOrderArr" :key="order.id">
            <div class="order-info">
              <p @click="toBusinessInfo(order.business.businessId)">
                {{ order.business.businessName }}
                <i class="fa fa-caret-right"></i>
              </p>
              <p>已支付</p>
            </div>
            <div class="order-detail">
              <div>
                <span v-for="item in order.food" :key="item.id">
                  <img :src="item.foodInfo.foodImg">
                </span>
              </div>
              <div>
                <p>&#165;{{ order.orderTotal }}</p>
                <p>共{{order.totalNum}}件</p>
              </div>
            </div>
            <div class="order-info-right">
              <div class="order-info-right-icon" @click="toBusinessInfo(order.business.businessId)">再来一单</div>
            </div>
          </li>
        </ul>
      <Footer></Footer>
    </div>
</template>
```

（2）在<script>标签中引入 axios 并创建 axios 实例，在 data 选项中定义登录用户信息、已支付订单列表和未支付订单列表。然后在 mounted 选项中使用 axios 发送 GET 请求，根据用户手机号码查找该用户的订单信息，再对查找的结果进行遍历。在遍历时通过判断订单状态将未支付订单加入未支付订单列表中，将已支付订单加入已支付订单列表中。接着分别定义 toBusinessInfo()方法和 toPay()方法，调用 toBusinessInfo()方法会跳转到商家详情页面，调用 toPay()方法会跳转到支付页面。代码如下：

```
<script>
  import Footer from "../components/TheFooter.vue";
  import axios from "axios";                              //引入 axios
  const api = axios.create({                              //创建 axios 实例
    baseURL: 'http://localhost:3000'
  });
  export default {
    name: "UserPayment",
    data() {
      return {
```

```
      user: JSON.parse(sessionStorage.getItem("user")),   //登录用户信息
      PaidOrderArr: [],                                    //已支付订单列表
      UnpaidOrderArr: []                                   //未支付订单列表
    }
  },
  components: {
    Footer
  },
  mounted() {
    //根据用户手机号码查找指定的订单信息
    api.get(`/OrderList?userId=${this.user.userId}`).then((response) => {
      let result = response.data;
      for (let orders of result) {
        if(orders.orderState === 0){
          this.UnpaidOrderArr.push(orders);                //将未支付订单加入未支付订单列表中
        }
        if(orders.orderState === 1){
          this.PaidOrderArr.push(orders);                  //将已支付订单加入已支付订单列表中
        }
      }
    }).catch((error) => {
      console.error(error);
    });
  },
  methods: {
    toBusinessInfo(businessId) {                           //跳转到商家详细信息页面
      this.$router.push({
        path: "/businessInfo",
        query: { businessId: businessId },
      });
    },
    toPay(id){                                             //跳转到支付页面
      this.$router.push({ path: "/payment", query: {orderId: id} })
    }
  }
}
</script>
```

8.10 注册和登录页面的设计与实现

在仿饿了么 APP 中，用户只有登录成功之后，才能将某商家的食物添加到购物车中，或者进入订单列表页面和我的页面。在登录之前，用户如果尚未注册，就需要先进行注册。用户注册页面和登录页面的效果分别如图 8.13 和图 8.14 所示。

图 8.13 用户注册页面

图 8.14 用户登录页面

8.10.1 注册页面的设计

在用户注册页面中,需要用户分别输入手机号码、密码、确认密码、用户姓名和性别,如果输入无误,单击"注册"按钮即可完成注册。具体实现步骤如下。

(1)在 views 文件夹下新建 UserRegister.vue 文件。在<template>标签中定义一个 ul 列表,在该列表中分别添加用于输入手机号码的文本框,用于输入密码和确认密码的文本框,用于输入姓名的文本框,用于选择性别的单选按钮。当触发"手机号码"文本框的 input 事件时,将调用 checkUserId()方法。在 ul 列表之后,添加一个"注册"按钮,当单击该按钮时,将调用 register()方法。代码如下:

```html
<template>
  <div class="wrapper">
    <header>
      <div>
        <i class="fa fa-angle-left" @click="$router.go(-1)"></i>
      </div>
      <p>用户注册</p>
    </header>
    <ul class="form-box">
      <li>
        <div class="title">手机号码:</div>
        <div class="content">
          <input type="text" @input="checkUserId" v-model="user.userId" placeholder="手机号码">
        </div>
      </li>
      <li>
        <div class="title">密码:</div>
        <div class="content">
          <input type="password" v-model="user.password" placeholder="密码">
        </div>
      </li>
      <li>
        <div class="title">确认密码:</div>
        <div class="content">
          <input type="password" v-model="confirmPassword" placeholder="确认密码">
        </div>
      </li>
      <li>
        <div class="title">用户姓名:</div>
        <div class="content">
          <input type="text" v-model="user.userName" placeholder="用户姓名">
        </div>
      </li>
      <li>
        <div class="title">性别:</div>
        <div class="content" style="font-size: 3.5vw">
          <input type="radio" v-model="user.userSex" value="1" style="width: 6vw; height: 3.2vw">男
          <input type="radio" v-model="user.userSex" value="0" style="width: 6vw; height: 3.2vw">女
        </div>
      </li>
    </ul>
    <div class="button-register">
      <button @click="register">注册</button>
    </div>
    <Footer></Footer>
  </div>
</template>
```

（2）在<script>标签中引入 axios 并创建 axios 实例，在 data 选项中定义用户信息对象和确认密码。然后定义 checkUserId()方法，该方法使用 axios 发送 GET 请求，根据用户输入的手机号码查找用户信息列表。如果查询结果不为空，就提示"该手机号码已存在"。接着定义 register()方法，该方法会检查用户输入的内容是否为空，如果所有输入都不为空，就使用 axios 发送 POST 请求，将注册信息添加到用户信息列表中。代码如下：

```vue
<script>
import Footer from "../components/TheFooter.vue";
import axios from "axios";                                    //引入 axios
const api = axios.create({                                    //创建 axios 实例
  baseURL: 'http://localhost:3000'
});
export default {
  name: "UserRegister",
  components: {
    Footer
  },
  data() {
    return {
      user: {
        userId: "",                                           //用户手机号码
        password: "",                                         //密码
        userName: "",                                         //用户姓名
        userSex: 1                                            //性别
      },
      confirmPassword: ""                                     //确认密码
    };
  },
  methods: {
    checkUserId(){
      if(this.user.userId !== ""){
        //根据手机号码查找用户
        api.get(`/UserList?userId=${this.user.userId}`).then((response) => {
          if (response.data.length > 0) {
            this.user.userId = "";
            alert("该手机号码已存在！");
          }
        }).catch((error) => {
          console.error(error);
        });
      }
    },
    register(){
      if (this.user.userId === "") {
        alert("手机号码不能为空！");
        return;
      }
      if (this.user.password === "") {
        alert("密码不能为空！");
        return;
      }
      if (this.user.password !== this.confirmPassword) {
        alert("两次输入的密码不一致！");
        return;
      }
      if (this.user.userName === "") {
        alert("用户姓名不能为空！");
        return;
      }
      //将用户信息添加到用户列表
```

```
        api.post("/UserList", this.user).then((response) => {
          if (response.data.id) {
            alert("注册成功！");
            this.$router.go(-1);
          } else {
            alert("注册失败！");
          }
        }).catch((error) => {
          console.error(error);
        });
      }
    }
  }
</script>
```

8.10.2 登录页面的设计

在用户登录页面中，用户需要分别输入手机号码和密码，如果输入正确，用户单击"登录"按钮即可成功登录。具体实现步骤如下。

（1）在 views 文件夹下新建 UserLogin.vue 文件。在<template>标签中定义一个 ul 列表，在该列表中分别添加用于输入手机号码的文本框和用于输入密码的文本框。在 ul 列表之后分别添加一个"登录"按钮和一个"去注册"按钮，当用户单击"登录"按钮时，将调用 login()方法；当用户单击"去注册"按钮时，将调用 register()方法。代码如下：

```
<template>
  <div class="wrapper">
    <header>
      <div>
        <i class="fa fa-angle-left" @click="$router.go(-1)"></i>
      </div>
      <p>用户登录</p>
    </header>
    <ul class="form-box">
      <li>
        <div class="title">手机号码：</div>
        <div class="content">
          <input type="text" v-model="userId" placeholder="手机号码">
        </div>
      </li>
      <li>
        <div class="title">密码：</div>
        <div class="content">
          <input type="password" v-model="password" placeholder="登录密码">
        </div>
      </li>
    </ul>
    <div class="button-login">
      <button @click="login">登录</button>
    </div>
    <div class="button-register">
      <button @click="register">去注册</button>
    </div>
    <Footer></Footer>
  </div>
</template>
```

（2）在<script>标签中引入 axios 并创建 axios 实例，在 data 选项中定义用户手机号码和登录密码。然后定义 login()方法，该方法首先会检查用户输入的内容是否为空。如果两者都不为空，就使用 axios 发送

GET 请求，根据输入的手机号码和密码查询用户信息列表。如果用户输入的手机号码和登录密码都正确，就将用户信息保存在 sessionStorage 中。接着定义 register()方法，调用该方法会跳转到用户注册页面。代码如下：

```html
<script>
  import Footer from "../components/TheFooter.vue";
  import axios from "axios";                                      //引入 axios
  const api = axios.create({                                      //创建 axios 实例
    baseURL: 'http://localhost:3000'
  });
  export default {
    name: "UserLogin",
    components: {
      Footer
    },
    data() {
      return {
        userId: "",                                               //用户手机号码
        password: "",                                             //密码
        prevName: ""                                              //前一个页面路由名称
      };
    },
    beforeRouteEnter(to, from, next){
      next(vm => {
        vm.prevName = from.name;
      })
    },
    methods: {
      login(){
        if (this.userId === "") {
          alert("手机号码不能为空！");
          return;
        }
        if (this.password === "") {
          alert("密码不能为空！");
          return;
        }
        //根据输入的手机号码和密码查询用户列表
        api.get(`/UserList?userId=${this.userId}&password=${this.password}`).then((response) => {
          let user = response.data[0];
          if (!user) {
            alert("用户名或密码不正确！");
          } else {
            //将用户信息保存在 sessionStorage 中
            sessionStorage.setItem("user", JSON.stringify(user));
            if(this.prevName){
              this.$router.go(-1);
            }else{
              this.$router.push("/index");                        //跳转到首页
            }        }
        }).catch((error) => {
            console.error(error);
        });
      },
      register(){
        this.$router.push({ path: "register" });                  //跳转到用户注册页面
      }
    }
  }
</script>
```

8.11 我的页面的设计与实现

在底部导航栏中,单击"我的"选项会跳转到我的页面。该页面主要展示当前登录用户的红包、卡券和地址等选项,其页面效果如图 8.15 所示。

具体实现步骤如下。

(1)在 views 文件夹下新建 MyHome.vue 文件。在<template>标签中定义两个 ul 列表,在这两个列表中分别添加当前登录用户的资产、红包、卡券和地址等选项。代码如下:

图 8.15 我的页面

```
<template>
  <div class="wrapper">
    <header>
      <div>
        <i class="fa fa-angle-left" @click="$router.go(-1)"></i>
      </div>
      <p>饿了么</p>
    </header>
    <div class="person">
      <span><img src="/images/home.png"></span>
      <span>{{user.userId}}</span>
    </div>
    <ul class="money">
      <li>
        <i class="fa fa-money"></i>
        <p>我的资产</p>
      </li>
      <li>
        <i class="fa fa-gift"></i>
        <p>红包</p>
      </li>
      <li>
        <i class="fa fa-credit-card"></i>
        <p>券包</p>
      </li>
    </ul>
    <ul class="item">
      <li>
        <i class="fa fa-map-marker"></i>
        <p>我的地址</p>
      </li>
      <li>
        <i class="fa fa-whatsapp"></i>
        <p>我的客服</p>
      </li>
      <li>
        <i class="fa fa-file-text"></i>
        <p>关于APP</p>
      </li>
      <li>
        <i class="fa fa-shield"></i>
        <p>用户隐私</p>
      </li>
      <li>
        <i class="fa fa-warning"></i>
```

```
        <p>规则中心</p>
      </li>
    </ul>
    <Footer></Footer>
  </div>
</template>
```

（2）在<script>标签中引入页面底部组件，在 data 选项中定义登录用户信息，在 components 选项中注册页面底部组件。代码如下：

```
<script>
  import Footer from "../components/TheFooter.vue";          //引入页面底部组件
  export default {
    name: "MyHome",
    data() {
      return {
        user: JSON.parse(sessionStorage.getItem("user"))     //登录用户信息
      };
    },
    components: {
      Footer
    }
  }
</script>
```

8.12 项目运行

通过前述步骤，我们已经设计并完成了"仿饿了么 APP"项目的开发。接下来，我们运行该项目，以检验我们的开发成果。首先打开 package.json 文件，在 scripts 配置选项中添加一个脚本命令，添加后的代码如下：

```
"scripts": {
  "serve": "vue-cli-service serve",
  "build": "vue-cli-service build",
  "lint": "vue-cli-service lint",
  "mock": "json-server data.json --port 3000"
}
```

打开命令提示符窗口，切换到项目目录，执行启动 JSON Server 的命令，如图 8.16 所示。

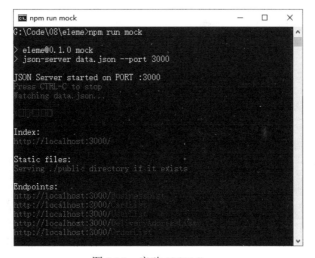

图 8.16 启动 JSON Server

新打开一个命令提示符窗口，切换到项目所在目录，执行 npm run serve 命令运行该项目，如图 8.17 所示。

在浏览器地址栏中输入 http://localhost:8080，然后按 Enter 键，即可成功运行该项目，页面效果如图 8.18 所示。

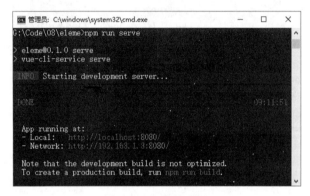

图 8.17 运行该项目

图 8.18 仿饿了么 APP 首页

8.13 源码下载

本章虽然详细地讲解了如何编码实现"仿饿了么 APP"的各个功能，但给出的代码都是代码片段，而非完整的源代码。为了方便读者学习，本书提供了该项目的完整源代码，读者可以扫描右侧的二维码进行下载。

第 9 章
仿今日头条 APP

——Vue CLI + Router + Vuex + axios + JSON Server + Vant + amfe-flexible + Day.js

随着互联网科技和新媒体的快速发展，人们获取新闻和信息的方式发生了很大的变化。作为一个新型的信息传播平台，今日头条以其个性化的新闻定制服务受到了广大用户的欢迎。本章将使用 Vue CLI、Router、Vuex、axios、JSON Server、Vant 等技术开发一个仿今日头条 APP 的移动端应用。

本项目的核心功能及实现技术如下：

项目微视频

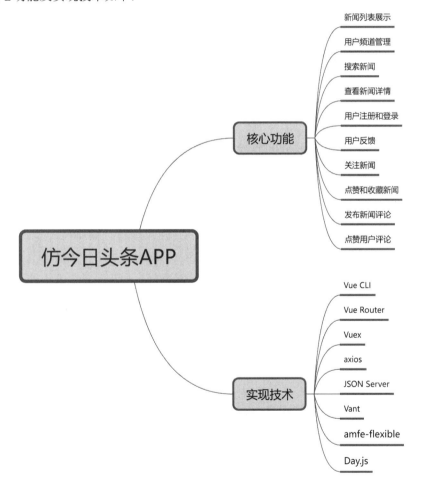

9.1 开发背景

随着网络设施的完善和智能手机的普及，互联网得到了迅速的发展，人们获取信息的方式已经从计算机端逐渐转移到了移动端，新闻类 APP 已经成为人们获取新闻资讯的主要渠道。在这一背景下，今日头条凭借其个性化推荐的模式成功突围，成为新闻类 APP 中的佼佼者。除了新闻资讯，今日头条还为用户提供了生活服务、实用工具、视频和直播等多种功能，以满足不同类型和需求的用户。本章将使用 Vue CLI、Router、Vuex、axios、JSON Server、Vant 等技术开发一个仿今日头条 APP 的移动端应用。其实现目标如下：

- ☑ 通过新闻列表页面展示新闻列表。
- ☑ 在频道管理弹出层中实现频道添加和删除功能。
- ☑ 在新闻详情页面展示新闻详细信息。
- ☑ 提供搜索新闻的功能。
- ☑ 提供用户注册和登录功能。
- ☑ 提供新闻反馈的功能。
- ☑ 提供关注新闻发布用户、点赞新闻和收藏新闻的功能。
- ☑ 提供发表评论和点赞用户评论的功能。

9.2 系统设计

9.2.1 开发环境

本项目的开发及运行环境如下：

- ☑ 操作系统：推荐 Windows 10、Windows 11 或更高版本，同时兼容 Windows 7（SP1）。
- ☑ 开发工具：WebStorm。
- ☑ 开发框架：Vue.js 3.0。

9.2.2 业务流程

在使用仿今日头条 APP 时，用户如果尚未注册或已经注册但是未登录，则只能浏览新闻列表、查看新闻详情、编辑用户频道和搜索新闻。用户只有登录后，才能进行关注新闻发布者、点赞新闻、收藏新闻、发表评论和点赞评论等操作。

根据该项目的业务需求，我们设计如图 9.1 所示的业务流程图。

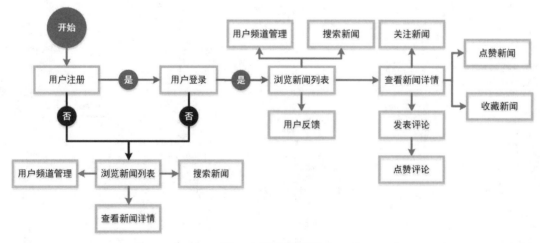

图 9.1 业务流程图

9.2.3 功能结构

本项目的功能结构已经在章首页中给出，其实现的具体功能如下：
- ☑ 新闻列表页面：展示不同频道下的新闻列表，包括新闻标题、新闻图片、新闻发布用户名、新闻评论数量以及新闻发布时间到当前时间的时间间隔等信息。
- ☑ 新闻搜索页面：在搜索框中输入搜索关键词，按 Enter 键或单击搜索框右侧的"搜索"按钮会跳转到新闻搜索结果页面，该页面将展示搜索到的新闻列表。
- ☑ 频道管理：展示用户频道列表和未选择频道列表。单击未选择频道列表中的某个频道可以将该频道添加到用户频道列表中；单击"编辑"文本后可以实现编辑用户频道的功能，单击用户频道列表中的某个频道可以将该频道从用户频道列表中删除，并将其添加到未选择频道列表中。
- ☑ 新闻详情页面：展示该新闻的详细信息，包括新闻标题、新闻发布用户头像、新闻发布用户名、新闻发布时间到当前时间的时间间隔、新闻内容和用户评论内容等信息。
- ☑ 注册和登录页面：为用户提供注册和登录的功能。用户只有在登录之后，才能对新闻进行反馈，以及关注新闻发布用户、点赞新闻、收藏新闻、发布评论和点赞评论的功能。
- ☑ 用户反馈：包括"不感兴趣"和"举报"两个功能。单击"不感兴趣"，系统将不再向当前登录用户展示该新闻，单击"举报"，则会展示对应的反馈列表。
- ☑ 关注新闻发布用户：在新闻详情页面，单击新闻标题下方的"关注"按钮可以关注该新闻的发布用户，再次单击按钮可以取消关注。
- ☑ 点赞和收藏新闻：在新闻详情页面，单击页面底部的心形图标，可以点赞或取消点赞该新闻，单击页面底部的星型图标，可以收藏或取消收藏该新闻。
- ☑ 发布评论：在新闻详情页面，单击页面底部的"发布评论"会显示输入框。在该输入框中填写评论内容后，单击"发布"按钮进行发布。发布后的评论内容会显示在文章内容的下方。
- ☑ 点赞评论：在评论列表中，单击某个评论右侧的心形图标，可以点赞或取消点赞该评论。

9.3 技术准备

9.3.1 技术概览

在开发仿今日头条 APP 时，我们主要应用 Vue CLI 脚手架工具、Vue Router、Vuex、axios、JSON Server、Vant、amfe-flexible 和 Day.js 等技术。其中，Vue CLI 脚手架工具、Vue Router、Vuex 和 axios 在《Vue.js 从入门到精通》中有详细的讲解，对这些知识不太熟悉的读者，可以参考该书对应的内容。JSON Server 已经在前面的章节做了简要介绍。下面将对 Vant、amfe-flexible 和 Day.js 进行必要介绍，以确保读者可以顺利完成本项目。

9.3.2 Vant

1. Vant 简介

Vant 是一个轻量、可靠的移动端组件库。Vant 能够帮助开发者快速搭建出风格统一的页面，从而提升开发效率。作为移动端组件库，Vant 一直将轻量化作为核心开发理念。为了平衡日益丰富的功能和轻量化之间的矛盾关系，Vant 内部使用了很多的优化方式，包括支持组件按需加载、公共模块复用、组件编译流程优化等。如果你的项目使用 Vue 3.0 进行开发，那么你需要使用 Vant 4.X 版本的组件库。

2．安装 Vant

要在现有项目中使用 Vant，可以通过 npm 方式进行安装。安装最新版 Vant 的命令如下：

```
npm install vant --save
```

3．注册组件

Vant 支持多种组件注册方式，可以根据实际业务需要进行选择。

1）全局注册

对 Vant 组件进行全局注册后，可以在项目中的任意子组件中使用注册的 Vant 组件。例如，全局注册按钮组件的代码如下：

```
import { createApp } from 'vue';
import { Button } from 'vant';
const app = createApp();
app.use(Button);
```

注册完成后，在模板中可以通过<van-button>标签来使用按钮组件。

2）全量注册

全量注册即一次性注册所有 Vant 组件。示例代码如下：

```
import Vant from 'vant';
import { createApp } from 'vue';
const app = createApp();
app.use(Vant);
```

> **注意**
> 使用全量注册会引入所有组件的代码，导致包的体积增大。

3）局部注册

局部注册即在当前组件中注册 Vant 组件。例如，在当前组件中注册按钮组件，代码如下：

```
import { Button } from 'vant';
export default {
    components: {
        [Button.name]: Button,
    },
};
```

4．Vant 常用组件

注册 Vant 组件后，就可以在项目中使用这些组件。下面对 Vant 中的一些常用组件进行介绍。

1）Button

Button 是按钮组件，使用按钮可以触发一个操作，如提交表单、执行某个函数等。Button 组件的常用属性及其说明如表 9.1 所示。

表 9.1 Button 组件的常用属性及其说明

属　性	说　　明	类　型	默　认　值
type	类型，可选值为 default、primary、success、warning、danger	string	default
size	尺寸，可选值为 normal、large、small、mini	string	normal
text	按钮文字	string	无默认值
color	按钮颜色，支持传入 linear-gradient 渐变色	string	无默认值

续表

属 性	说 明	类 型	默 认 值
icon	左侧图标名称或图片链接，等同于 Icon 组件的 name 属性	string	无默认值
round	是否为圆形按钮	boolean	false
disabled	是否禁用按钮	boolean	false
loading	是否显示为加载状态	boolean	false
loading-text	加载状态提示文字	string	无默认值
to	单击后跳转的目标路由对象，等同于 Vue Router 的 to 属性	string | object	无默认值

例如，在页面中添加一个小型的警告按钮，代码如下：

```
<van-button type="warning" size="small">警告按钮</van-button>
```

2）NavBar

NavBar 是导航栏组件，它可以为页面提供导航功能，常用于页面顶部。NavBar 组件的常用属性及其说明如表 9.2 所示。

表 9.2　NavBar 组件的常用属性及其说明

属 性	说 明	类 型	默 认 值
title	标题	string	无默认值
left-text	左侧文字	string	无默认值
right-text	右侧文字	string	无默认值
left-arrow	是否显示左侧箭头	boolean	false
border	是否显示下边框	boolean	true
fixed	是否固定在顶部	boolean	false

例如，设置导航栏标题为"用户注册"，显示左侧箭头，当单击左侧箭头时返回上一页，代码如下：

```
<van-nav-bar title="用户注册" left-arrow @click-left="$router.go(-1)" />
```

3）Tabs 和 Tab

Tabs 和 Tab 是选项卡组件，使用这两个组件可以在不同的内容区域之间进行切换。Tabs 组件的常用属性及其说明如表 9.3 所示。

表 9.3　Tabs 组件的常用属性及其说明

属 性	说 明	类 型	默 认 值
v-model:active	绑定当前选中标签的标识符	number | string	0
type	样式风格类型，可选值为 card	string	line
color	标签主题色	string	#1989fa
background	标签栏背景色	string	white
animated	是否开启切换标签内容时的转场动画	boolean	false
sticky	是否使用粘性布局	boolean	false

Tab 组件的常用属性及其说明如表 9.4 所示。

表 9.4　Tab 组件的常用属性及其说明

属 性	说 明	类 型	默 认 值
title	标题	string	无默认值
disabled	是否禁用标签	boolean	false

续表

属　性	说　明	类　型	默　认　值
dot	是否在标题右上角显示小红点	boolean	false
badge	图标右上角徽标的内容	number | string	无默认值
name	标签名称，作为匹配的标识符	number | string	标签的索引值
to	单击后跳转的目标路由对象，等同于 Vue Router 的 to 属性	string | object	无默认值

例如，设置一组选项卡，通过 v-model:active 绑定当前激活标签对应的索引值，默认情况下启用第一个标签，并开启切换标签内容时的转场动画，代码如下：

```
<template>
    <van-tabs v-model:active="active" animated>
        <van-tab title="电影">电影内容</van-tab>
        <van-tab title="音乐">音乐内容</van-tab>
        <van-tab title="动漫">动漫内容</van-tab>
    </van-tabs>
</template>
<script>
export default {
  data() {
    return {
      active: 0
    }
  }
}
</script>
```

4）Icon

Icon 是图标组件，Icon 组件允许开发者使用基于字体的图标集，同时可以在其他组件中通过 icon 属性使用图标。Icon 组件的常用属性及其说明如表 9.5 所示。

表 9.5　Icon 组件的常用属性及其说明

属　性	说　明	类　型	默　认　值
name	图标名称或图片链接	string	无默认值
dot	是否显示图标右上角小红点	boolean	false
badge	图标右上角徽标的内容	number | string	无默认值
color	图标颜色	string	inherit
size	图标大小，如 20px 2em，默认单位为 px	number | string	inherit
class-prefix	类名前缀，用于使用自定义图标	string	van-icon

例如，设置一个表示删除的图标，并设置图标的颜色和大小，代码如下：

```
<van-icon name="cross" color="#666666" size="0.3rem"/>
```

5）Popup

Popup 是弹出层组件，用于展示弹窗、信息提示等内容，并支持多个弹出层叠加展示。Popup 组件的常用属性及其说明如表 9.6 所示。

表 9.6　Popup 组件的常用属性及其说明

属　性	说　明	类　型	默　认　值
v-model:show	是否显示弹出层	boolean	false
overlay	是否显示遮罩层	boolean	true

续表

属性	说明	类型	默认值
position	弹出位置，可选值为 center、top、bottom、right、left	string	center
round	是否显示圆角	boolean	false
closeable	是否显示关闭图标	boolean	false
teleport	指定挂载的节点，等同于 Teleport 组件的 to 属性	string | Element	无默认值

例如，要设置一个弹出层，并显示其圆角，同时将该弹出层挂载到 body 节点下，可以使用以下代码：

```
<van-popup v-model:show="show" round teleport="body">弹出层内容</van-popup>
```

6）Row 和 Col

Row 和 Col 组件用来进行行列布局。这两个组件共同构成了一个 24 列的栅格系统，你可以通过在 Col 组件上添加 span 属性设置列宽所占的百分比。此外，你还可以添加 offset 属性来设置列的偏移宽度，其计算方式与 span 相同。Row 组件的常用属性及其说明如表 9.7 所示。

表 9.7 Row 组件的常用属性及其说明

属性	说明	类型	默认值
gutter	列元素之间的间距（单位为 px）	number | string	无默认值
tag	自定义元素标签	string	div
justify	主轴对齐方式，可选值为 start、end、center、space-around、space-between	string	start
align	交叉轴对齐方式，可选值为 top、center、bottom	string	top
wrap	是否自动换行	boolean	true

Col 组件的常用属性及其说明如表 9.8 所示。

表 9.8 Col 组件的常用属性及其说明

属性	说明	类型	默认值
span	列元素宽度	number | string	无默认值
offset	列元素偏移距离	number | string	无默认值
tag	自定义元素标签	string	div

例如，使用 Row 组件的 justify 属性设置主轴上每个元素的两侧间隔相等，代码如下：

```
<van-row justify="space-around">
    <van-col span="6">视频</van-col>
    <van-col span="6">直播</van-col>
    <van-col span="6">养生</van-col>
</van-row>
```

7）Cell

Cell 是单元格组件，它可以作为列表中的单个展示项。Cell 组件的常用属性及其说明如表 9.9 所示。

表 9.9 Cell 组件的常用属性及其说明

属性	说明	类型	默认值
title	左侧标题	number | string	无默认值
value	右侧内容	number | string	无默认值
label	标题下方的描述信息	string	无默认值
size	单元格大小，可选值为 large	string	无默认值

续表

属性	说明	类型	默认值
icon	左侧图标名称或图片链接，等同于 Icon 组件的 name 属性	string	无默认值
clickable	是否开启单击反馈	boolean	null
is-link	是否展示右侧箭头并开启单击反馈	boolean	false
arrow-direction	箭头方向，可选值为 left、right、up、down	string	right

例如，设置一个单元格，通过 is-link 属性在单元格右侧显示箭头，并通过 arrow-direction 属性控制箭头方向，代码如下：

```
<van-cell title="单元格" is-link arrow-direction="down" value="内容" />
```

8）List

List 是列表组件，它主要用于展示长列表，当列表即将滚动到页面底部时会触发事件并加载更多列表项。List 组件的常用属性及其说明如表 9.10 所示。

表 9.10 List 组件的常用属性及其说明

属性	说明	类型	默认值
v-model:loading	是否处于加载状态，加载过程中不触发 load 事件	boolean	false
finished	是否已加载完成，加载完成后不再触发 load 事件	boolean	false
offset	滚动条与页面底部距离小于 offset 时，触发 load 事件	number | string	300
loading-text	加载过程中的提示文字	string	加载中…
finished-text	加载完成后的提示文字	string	无默认值
immediate-check	是否在初始化时立即执行滚动位置检查	boolean	true
disabled	是否禁用滚动加载	boolean	false
direction	滚动触发加载的方向，可选值为 up	string	down

List 组件通过 loading 和 finished 两个变量控制加载状态，当组件滚动到页面底部时，会触发 load 事件并将 loading 设置为 true。此时可以发起异步操作并更新数据，数据更新完毕后，将 loading 设置为 false 即可。若数据已全部加载完毕，则直接将 finished 设置为 true 即可。

例如，为列表组件设置加载状态，如果数据已加载完成，就显示相应的提示文字，代码如下：

```
<van-list
  v-model:loading="loading"
  :finished="finished"
  finished-text="没有更多了"
  @load="onLoad"
>
  长列表内容
</van-list>
```

9）Divider

Divider 是分割线组件，它用于将内容分隔为多个区域。Divider 组件的常用属性及其说明如表 9.11 所示。

表 9.11 Divider 组件的常用属性及其说明

属性	说明	类型	默认值
dashed	是否使用虚线	boolean	false
hairline	是否使用 0.5px 线	boolean	true
content-position	内容位置，可选值为 center、left、right	string	center

例如，在页面中添加一条分割线，并设置分割线中间的文本，代码如下：

```
<van-divider>End</van-divider>
```

9.3.3　amfe-flexible

amfe-flexible 是一个用于移动端适配的 JavaScript 库，它可以根据设备的屏幕尺寸动态地计算出所需的 rem 基准值，从而方便开发者使用 rem 单位进行页面布局和样式定义。要使用 amfe-flexible 进行移动端适配，需要先安装该库，然后在 main.js 文件中引入它。安装 amfe-flexible 插件的命令如下：

```
npm install amfe-flexible --save
```

安装完成后，需要在 main.js 文件中引入 amfe-flexible，引入的代码如下：

```
import 'amfe-flexible';
```

在安装和引入 amfe-flexible 后，amfe-flexible 会根据设备的屏幕宽度和基准字体大小自动计算出实际的像素值，并将其应用到页面元素上。

9.3.4　Day.js

Day.js 是一个轻量的处理日期和时间的 JavaScript 库。与一些庞大的日期库不同，Day.js 的设计理念是保持简洁、易用和灵活。它的大小仅有 2 KB，而且拥有强大的功能。Day.js 提供了简单而直观的 API，使得日期和时间的操作变得非常容易。

要使用 Day.js 库处理日期和时间，需要先安装该库，然后在代码中引入它。安装 Day.js 库的命令如下：

```
npm install dayjs --save
```

安装完成后，需要在代码中引入 Day.js，引入的代码如下：

```
import dayjs from "dayjs";
```

在安装并引入 Day.js 后，就可以使用其 API 来处理日期和时间。在该项目中，展示新闻和评论的发布时间都使用相对时间，下面对如何使用 Day.js 获取相对时间进行简单介绍。

要使用 Day.js 获取相对时间，需要导入 relativeTime 插件。另外，Day.js 使用的默认语言是英语，如果要显示中文的相对时间，需要手动导入中文语言包，将一些描述性的文字转换为中文进行显示。获取相对时间的代码如下：

```
import dayjs from "dayjs";
import relativeTime from "dayjs/plugin/relativeTime";
import "dayjs/locale/zh-cn";
export const timeInterval = (targetTime) => {
  dayjs.extend(relativeTime);
  dayjs.locale("zh-cn");
  var a = dayjs();
  var b = dayjs(targetTime);
  return a.to(b);
};
```

> **说明**
>
> 在安装 Day.js 时，会同时安装语言包和插件包。语言包位于 dayjs/locale 目录下，而插件包则位于 dayjs/plugin 目录下。

9.4 创建项目

在设计仿今日头条 APP 各功能模块之前,需要使用 Vue CLI 创建项目,并将项目名称设置为 headlines。在命令提示符窗口中,首先切换到要创建的项目所在的目录,然后输入以下命令:

```
vue create headlines
```

按 Enter 键,选择 Manually select features,如图 9.2 所示。

按 Enter 键后,选择 Router、Vuex 和 CSS Pre-processors 选项,如图 9.3 所示。

图 9.2　选择 Manually select features

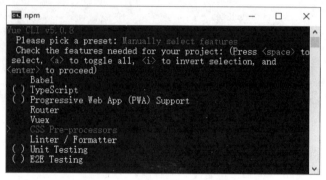

图 9.3　选择配置选项

按 Enter 键后,选择 Vue 3.x 版本,如图 9.4 所示。

然后选择路由是否使用 history 模式,输入 y 表示使用 history 模式,如图 9.5 所示。

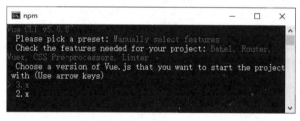

图 9.4　选择 Vue 版本

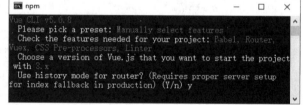

图 9.5　使用 history 模式

接下来选择一个 CSS 预处理器,选择 Less 选项,如图 9.6 所示。

在选择配置信息的存放位置时选择 In package.json 选项,即将配置信息存储在 package.json 文件中,如图 9.7 所示。

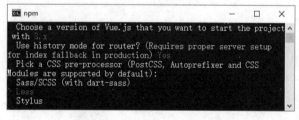

图 9.6　选择 CSS 预处理器

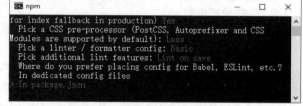

图 9.7　选择配置信息的存放位置

创建项目后进入项目目录。由于该项目使用了最新版的 Vant 组件库,因此为了实现 Vue 和 Vant 的兼容，需要安装最新版的 Vue.js。安装命令如下：

npm install vue --save

然后分别安装项目需要使用的插件和库，包括 axios、Vant、amfe-flexible 和 Day.js。

安装后整理项目目录，在 public 目录中创建 images 文件夹，在该文件夹中存储项目需要用到的图片文件。

此外，由于该项目使用了 JSON Server 模拟后端服务器，因此还需要安装 JSON Server，在项目根目录下创建一个 data.json 文件作为数据源，在该文件中添加模拟服务器的数据。

9.5 新闻列表页面的设计与实现

新闻列表页面主要由项目名称、用户频道列表、新闻列表和底部导航栏组成。新闻列表页面的效果如图 9.8 所示。

新闻列表页面主要涉及 5 个组件，分别是页面主组件、新闻列表组件、新闻列表项组件、频道管理组件和底部导航栏组件，下面对这 5 个组件进行详细讲解。

9.5.1 页面主组件设计

新闻列表页面的主组件是新闻列表组件和频道管理组件的父组件，该组件主要用来设置项目名称、遍历用户频道列表，以及在合适的位置引入新闻列表组件和频道管理组件。新闻列表页面主组件的主要实现步骤如下。

（1）在 views 文件夹下新建 Home 文件夹，在 Home 文件夹下新建 index.vue 文件。在<template>标签中：使用 NavBar 组件定义导航栏；使用 Tabs 和 Tab 组件定义用户频道列表，在 Tab 组件中调用新闻列表组件 NewsList.vue；使用 Icon 组件定义图标，单击该图标调用 wapNavClick()方法；使用 Popup 组件定义弹出层，在弹出层中调用频道管理组件 ChannelEdit.vue。代码如下：

图 9.8 仿今日头条新闻列表页面

```
<template>
  <div>
    <div>
      <van-nav-bar title="今日头条" fixed>
        <template #right>
          <van-icon
            name="search"
            size="0.52rem"
            @click="moveSearchPage"
          />
        </template>
      </van-nav-bar>
    </div>
    <div class="main">
      <van-tabs v-model:active="channelId" sticky offset-top="1.23rem"
             color="#FF0000" title-active-color="#FF0000" line-height="2" animated>
        <van-tab
```

```html
          v-for="obj in userChannelList"
          :title="obj.name"
          :key="obj.id"
          :name="obj.id"
        >
          <NewsList :channelId="channelId"></NewsList>
        </van-tab>
      </van-tabs>
      <van-icon
        name="wap-nav"
        size="0.4rem"
        class="moreChannels"
        @click="wapNavClick"
      ></van-icon>
    </div>
    <van-popup v-model:show="show" round
               teleport="body" style="width: 100%; height: 100%;">
      <ChannelEdit
        :userList="userChannelList"
        :unCheckList="unCheckChannelList"
        @addChannelEV="addChannel"
        @removeChannel="removeChannel"
        @close="closeFun"
        @toggle="toggleChannel"
      ></ChannelEdit>
    </van-popup>
  </div>
</template>
```

（2）在<script>标签中分别引入 axios 实例、新闻列表组件和频道管理组件。在 data 选项中定义数据；在 created()钩子函数中分别查询用户频道列表和所有频道列表；在 methods 选项中定义方法。其中，wapNavClick()方法用于显示频道管理弹出层，addChannel()方法用于将单击的频道添加到用户频道列表中，removeChannel()方法用于从用户频道列表中删除该频道，closeFun()方法用于关闭弹出层，moveSearchPage()方法用于跳转到新闻搜索页面，toggleChannel()方法用于切换频道。最后定义计算属性 unCheckChannelList，该计算属性用于获取用户未选择的频道列表。代码如下：

```javascript
<script>
import api from "@/utils/request";                      //引入 axios 实例
import NewsList from "./NewsList";                       //引入新闻列表组件
import ChannelEdit from "./ChannelEdit.vue";             //引入频道管理组件
export default {
  name: "HomeIndex",
  components: { NewsList, ChannelEdit },
  data() {
    return {
      channelId: 0,                                      //当前频道 id
      userChannelList: [],                               //用户频道列表
      allChannelList: [],                                //所有频道列表
      NewsList: [],                                      //新闻列表
      show: false                                        //控制是否显示频道编辑弹出层
    };
  },
  async created() {
    //查询用户频道列表
    await api({
      url: '/channels',
      method: 'GET'
    }).then((res) => {
      this.userChannelList = res.data;
    });
```

```javascript
      //查询所有频道列表
      await api({
        url: '/AllChannels',
        method: 'GET'
      }).then((res) => {
        this.allChannelList = res.data;
      });
    },
    methods: {
      wapNavClick() {                                              //显示频道管理弹出层
        this.show = true;
      },
      async addChannel(channelObj) {                               //将该频道添加到用户频道列表中
        await api({
          url: "/channels",
          method: "POST",
          data: {
            id: channelObj.id,
            name: channelObj.name
          }
        })
        //更新用户频道列表
        this.userChannelList.push(channelObj);
      },
      async removeChannel(channelObj) {                            //删除频道
        //获取要删除频道的索引
        const index = this.userChannelList.findIndex((obj) => {
          return obj.id === channelObj.id;
        });
        this.userChannelList.splice(index, 1);                     //从用户频道列表中删除该频道
        //执行删除操作
        await api({
          url: `/channels/${channelObj.id}`,
          method: "DELETE"
        })
      },
      closeFun() {                                                 //关闭弹出层
        this.show = false;
      },
      moveSearchPage() {
        this.$router.push("/search");                              //跳转到新闻搜索页面
      },
      toggleChannel(channelId){                                    //切换频道
        this.channelId = channelId;
      }
    },
    computed: {
      unCheckChannelList() {
        const newArr = [];                                         //用户未选择的频道列表
        //获取用户频道名称组成的数组
        let nameArr = this.userChannelList.map(smallObj => smallObj.name);
        this.allChannelList.forEach((bigObj) => {
          //如果 nameArr 数组中不包括遍历的频道名称，就把该频道添加到 newArr 数组中
          if(!nameArr.includes(bigObj.name)){
            newArr.push(bigObj);
          }
        });
        return newArr;
      },
    }
  }
};
</script>
```

9.5.2 新闻列表组件设计

新闻列表组件是新闻列表页面主组件的子组件，该组件主要用来遍历新闻列表，并实现用户反馈的功能。该组件使用了一个关键的技术，就是通过 Vant 中的 List 组件实现上拉加载功能，以及通过 PullRefresh 组件实现下拉刷新的功能。新闻列表组件的主要实现步骤如下。

（1）在 Home 文件夹下新建 NewsList.vue 文件。在<template>标签中，使用 List 组件和 PullRefresh 组件实现上拉加载和下拉刷新功能。在 List 组件中，调用新闻列表项组件 NewsItem.vue，在调用组件时对查询到的新闻列表数组进行遍历，并将遍历获得的每个新闻对象作为 Prop 属性的值传递给 NewsItem.vue 组件。当触发自定义事件 dislike 时，调用 dislikeFun()方法；当触发自定义事件 report 时，调用 reportFun()方法；当单击组件时，调用 itemClick()方法。最后定义弹出层，当 show 属性值为 true 时显示该弹出层，在弹出层中显示相应的提示信息。代码如下：

```
<template>
  <div>
    <van-pull-refresh v-model="refreshing" @refresh="onRefresh">
      <van-list
        v-model:loading="loading"
        :finished="finished"
        finished-text="没有更多了"
        @load="onLoad"
      >
        <NewsItem
          v-for="obj in newList"
          :key="obj.id"
          :newsObj="obj"
          @dislike="dislikeFun"
          @report="reportFun"
          @click="itemClick(obj.id)"
        ></NewsItem>
      </van-list>
    </van-pull-refresh>
    <van-popup v-model:show="show"
      closeable
      round
      teleport="body"
      style="width: 60%; height: 30%; text-align: center"
    >
      <van-cell>
        <view class="cell-text">登录后可反馈</view>
      </van-cell>
      <van-button style="width: 80%;" type="danger" to="/login">一键登录</van-button>
    </van-popup>
  </div>
</template>
```

（2）在<script>标签中，分别引入 axios 实例、新闻列表项组件和 Vant 中的轻提示组件 showToast，然后分别定义传递的 Prop 属性和注册的组件。在 data 选项中定义数据；在 created()钩子函数中判断用户是否已登录，如果已登录，就分别查询 JSON 服务器中的 dislike 和 feedback 数据接口，将用户不感兴趣的新闻 id 和举报的新闻 id 添加到数组中；在 computed 选项中定义计算属性 newList，该计算属性用于过滤 list 数组；在 methods 选项中定义方法。其中：onRefresh()方法用于下拉刷新数据；onLoad()方法用于上拉加载数据；dislikeFun()方法用于将用户反馈不感兴趣的文章 id 添加到 JSON 服务器上；reportFun()方法用于将用户举报的文章 id 添加到 JSON 服务器上；itemClick()方法用于跳转到新闻详情页面，并在该方法中将新闻 id 作为

参数进行传递。代码如下:

```html
<script>
import api from "@/utils/request";                              //引入 axios 实例
import NewsItem from "../../components/NewsItem.vue";           //引入新闻列表项组件
import { showToast } from "vant";                               //引入 Vant 中的轻提示组件
export default {
  props: {
    channelId: Number,
  },
  components: {
    NewsItem,
  },
  data() {
    return {
      list: [],                                                 //新闻列表数组
      loading: false,                                           //是否处于加载状态
      finished: false,                                          //是否已加载完成
      theTime: new Date().getTime(),                            //当前时间戳
      refreshing: false,                                        //是否处于刷新状态
      now: 0,                                                   //第几页
      limit: 10,                                                //每页显示数据个数
      totalNumber: 0,                                           //该频道下的新闻总数量
      NewsIdArr: [],                                            //已反馈的新闻 id 数组
      show: false                                               //是否显示弹出层
    };
  },
  async created() {
    if(localStorage.getItem("userId")){
      await api({
        url: "/dislike",
        method: "GET",
        params: {
          userId: localStorage.getItem("userId")
        }
      }).then((res) => {
        res.data.forEach((item) => {
          this.NewsIdArr.push(item.news_id);                    //将不感兴趣的新闻 id 添加到数组中
        })
      })
      await api({
        url: "/feedback",
        method: "GET",
        params: {
          userId: localStorage.getItem("userId")
        }
      }).then((res) => {
        res.data.forEach((item) => {
          this.NewsIdArr.push(item.news_id);                    //将举报的新闻 id 添加到数组中
        })
      })
    }
  },
  computed: {
    newList(){                                                  //过滤后显示在页面中的新闻列表
      return this.list.filter((item) => {
        return !this.NewsIdArr.includes(item.id);
      })
    }
  },
```

```js
methods: {
  async onRefresh() {                                         //下拉刷新
    this.list = [];                                           //清空列表数据
    this.refreshing = false;
    this.finished = false;
    this.loading = true;
    this.now = 0;
    await this.onLoad();
  },
  async onLoad() {                                            //上拉加载
    await api({
      url: '/NewsList',
      method: 'GET',
      params: {
        channel_id: this.channelId
      }
    }).then((res) => {
      this.loading = true;
      this.totalNumber = res.data.length;                     //获取该频道下的新闻总数量
      if(this.totalNumber <= this.limit){
        this.finished = true;                                 //加载完成
      }
    });
    this.now += 1;                                            //将当前页数加1
    //分页查询新闻列表
    await api({
      url: '/NewsList',
      method: 'GET',
      params: {
        channel_id: this.channelId,
        _page: this.now,
        _per_page: this.limit
      }
    }).then((res) => {
      this.loading = false;
      this.list = [...this.list, ...res.data.data]
      //数据全部加载完成
      if (this.list.length >= this.totalNumber) {
        this.finished = true;
      }
    });
  },
  //反馈不感兴趣
  async dislikeFun(newsObj) {
    if(!localStorage.getItem("userId")){
      this.show = true;
    }else {
      await api({
        url: "/dislike",
        method: "POST",
        data: {
          userId: localStorage.getItem("userId"),
          news_id: newsObj.id
        }
      }).then(() => {
        showToast("将减少此类内容推荐");
        this.list = this.list.filter(item => {
          return item !== newsObj;
        })
      })
```

```
      }
    },
    //反馈举报
    async reportFun(newsObj, type) {
      if(!localStorage.getItem("userId")){
        this.show = true;
      }else{
        await api({
          url: "/feedback",
          method: "POST",
          data: {
            userId: localStorage.getItem("userId"),
            news_id: newsObj.id,
            type: type
          }
        }).then(() => {
          showToast("举报提交成功，感谢反馈");
          this.list = this.list.filter(item => {
            return item !== newsObj;
          })
        })
      }
    },
    itemClick(id) {                                              //进入新闻详情页面
      this.$router.push({
        path: '/detail',
        query: {news_id: id}
      });
    },
  },
};
</script>
```

9.5.3 新闻列表项组件设计

新闻列表项组件是新闻列表组件的子组件，该组件主要用来显示新闻列表中每条新闻的信息，包括新闻标题、新闻图片、新闻发布用户名、新闻评论数量和新闻发布时间到当前时间的时间间隔等信息。另外，在该组件中还提供了用户反馈功能，每条新闻的右下角都有一个删除图标，单击该图标会显示用户反馈面板的内容。反馈面板中的选项分为一级反馈和二级反馈，其中一级反馈的界面如图 9.9 所示。单击一级反馈面板中的"举报"选项打开二级反馈面板，界面如图 9.10 所示。

图 9.9　一级反馈面板　　　　　　图 9.10　二级反馈面板

新闻列表项组件的主要实现步骤如下。

（1）在 components 文件夹下新建 NewsItem.vue 文件。在<template>标签中使用 Vant 中的 Cell 组件定义新闻列表的内容。在 Cell 组件中添加两个插槽：title 和 label。其中，title 插槽用于定义新闻标题和新闻图片，label 插槽用于定义新闻发布用户名、新闻评论数量和新闻发布时间到当前时间的时间间隔。在 label 插槽中还添加了一个删除图标，单击该图标将 show 属性的值设置为 true，这时会显示下面的动作面板组件 ActionSheet，在该面板中展示用户反馈的一些选项。当选择一级反馈面板或二级反馈面板中的某个选项时，调用 onSelect()方法。当单击一级反馈面板中的"取消"或二级反馈面板中的"返回"选项时，调用 cancelFun()方法。当单击反馈面板以外的区域时，会关闭该面板，此时会调用 closeFun()方法。代码如下：

```
<template>
  <van-cell>
    <template #title>
      <div class="title-box">
        <span>{{ newsObj.title }}</span>
        <img
          v-if="newsObj.cover.number === 1"
          class="thumb"
          :src="newsObj.cover.images[0]"
          alt=""
        />
      </div>
      <div class="thumb-box" v-if="newsObj.cover.number > 1">
        <img
          v-for="(imgUrl, index) in newsObj.cover.images"
          :key="index" class="thumb" :src="imgUrl" alt=""
        />
      </div>
    </template>
    <template #label>
      <div class="label-box">
        <div>
          <span>{{ newsObj.aut_name }}</span>
          <span>{{ newsObj.comm_count }}评论</span>
          <span>{{ timeInterval(newsObj.pubDate) }}</span>
        </div>
        <van-icon name="cross" @click.stop="show = true"/>
        <van-action-sheet
          v-model:show="show"
          :actions="actions"
          @select="onSelect"
          @cancel="cancelFun"
          @close="closeFun"
          teleport="body"
          :cancel-text="bottomText"
        />
      </div>
    </template>
  </van-cell>
</template>
```

说明

当单击每条新闻右下角的×图标时，为了防止事件冒泡，需要使用.stop 事件修饰符，否则单击×图标会直接跳转到新闻详情页面。

（2）在 src 文件夹下新建 utils 文件夹，在 utils 文件夹中定义 date.js 文件，在该文件中编写获取当前时间到指定时间的时间间隔的函数 timeInterval()，代码如下：

```
import dayjs from "dayjs";
import relativeTime from "dayjs/plugin/relativeTime";
import "dayjs/locale/zh-cn";
export const timeInterval = (targetTime) => {
  dayjs.extend(relativeTime);
  dayjs.locale("zh-cn");
  var a = dayjs();
  var b = dayjs(targetTime);
  return a.to(b);
};
```

（3）打开 store/index.js 文件，在 state 对象中定义用户反馈面板的选项内容，代码如下：

```
import { createStore } from 'vuex'
export default createStore({
  state: {
    firstActions : [
      { name: '不感兴趣' },
      { name: '举报' }
    ],
    secondActions : [
      {
        value: 0,
        name: '其他问题'
      },
      {
        value: 1,
        name: '标题夸张'
      },
      {
        value: 2,
        name: '旧闻重复'
      },
      {
        value: 3,
        name: '封面反感'
      },
      {
        value: 4,
        name: '内容质量差'
      }
    ]
  }
});
```

（4）在<script>标签中分别引入 timeInterval()函数和 mapState()辅助函数，然后定义传递的 Prop 属性。在 data 选项中定义数据，在 computed 选项中实现组件中的计算属性和 state 对象的数据之间的映射，在 created()钩子函数中将 actions 数组的值定义为一级反馈面板的内容，在 methods 选项中注册 timeInterval()函数和定义方法。其中，onSelect()方法用于对用户反馈的选项进行处理，cancelFun()方法用于返回一级反馈面板或取消用户反馈，closeFun()方法用于执行关闭用户反馈面板后的操作。代码如下：

```
<script>
import { timeInterval } from "@/utils/date.js";           //引入时间间隔函数
import {mapState} from "vuex";                            //引入 mapState 辅助函数
export default {
  props: {
    newsObj: Object,
  },
  data() {
```

```js
      return {
        show: false,              //控制用户反馈面板的显示状态
        actions: [],              //用户反馈列表
        bottomText: "取消",       //用户反馈面板底部的文字
      };
    },
    computed: {
      ...mapState([
        'firstActions',
        'secondActions'
      ])
    },
    created() {
      this.actions = this.firstActions;
    },
    methods: {
      timeInterval,                //时间间隔函数
      onSelect(item) {
        if (item.name === "举报") {
          this.actions = this.secondActions;
          this.bottomText = "返回";
        } else if (item.name === "不感兴趣") {
          //子组件向父组件传递新闻对象，在父组件中执行不感兴趣的用户反馈
          this.$emit("dislike", this.newsObj);
          this.show = false;
        } else {
          //子组件向父组件传递新闻对象和举报分类，在父组件中执行举报的用户反馈
          this.$emit("report", this.newsObj, item.value);
          this.show = false;
        }
      },
      //返回一级反馈面板或取消用户反馈
      cancelFun() {
        if (this.bottomText === "返回") {
          this.show = true;
          this.actions = this.firstActions;
          this.bottomText = "取消";
        } else {
          this.show = false;
        }
      },
      //关闭用户反馈面板后执行的操作
      closeFun() {
        this.actions = this.firstActions;
        this.bottomText = "取消";
      },
    },
  };
</script>
```

9.5.4 频道管理组件设计

频道管理组件是新闻列表页面主组件的子组件，该组件主要用来管理用户频道。在频道管理弹出层中分别展示用户已选择的频道列表（我的频道列表）和未选择的热门精选频道列表，界面效果如图 9.11 所示。单击未选择的热门精选频道列表中的某个频道，该频道会被移动到我的频道列表中，效果如图 9.12 所示。单击界面中的"编辑"按钮进入我的频道可编辑的状态，此时单击我的频道中的某个频道，该频道会被移动到未选择的热门精选频道列表中，效果如图 9.13 所示。

图 9.11 频道管理界面　　　图 9.12 移动到我的频道列表中　　　图 9.13 移动到热门精选频道列表中

频道管理组件的主要实现步骤如下。

（1）在 Home 文件夹下新建 ChannelEdit.vue 文件。在<template>标签中分别定义导航栏组件和两个<div>标签。在第一个<div>标签中定义我的频道列表，单击某个频道时，调用 userChannelClick()方法；在第二个<div>标签中定义用户未选择的热门精选频道列表，单击某个频道时，调用 more()方法。代码如下：

```
<template>
  <div>
    <van-nav-bar title="频道管理">
      <template #right>
        <van-icon
          name="cross"
          size="0.4rem"
          @click="close"
        />
      </template>
    </van-nav-bar>
    <div class="my-channel-box">
      <div class="channel-title">
        <span>我的频道
          <span class="small-title">
            点击{{ isEdit ? "删除" : "进入" }}频道
          </span>
        </span>
        <span class="channel-action" @click="edit">{{ isEdit ? "完成" : "编辑" }}</span>
      </div>
      <van-row>
        <van-col
          span="6"
          v-for="obj in userList"
          :key="obj.id"
```

```
            @click="userChannelClick(obj)"
          >
            <div class="channel-item van-hairline--surround">
              {{ obj.name }}
              <van-badge
                color="transparent"
                class="cross-badge"
                v-show="isEdit && obj.id !== '0'"
              >
                <template #content>
                  <van-icon name="cross" color="#666666" size="0.26rem"/>
                </template>
              </van-badge>
            </div>
          </van-col>
        </van-row>
      </div>
      <div class="more-channel-box">
        <div class="channel-title">
          <span>热门精选频道</span>
        </div>
        <van-row>
          <van-col
            span="6"
            v-for="obj in unCheckList"
            :key="obj.id"
            @click="more(obj)"
          >
            <div class="channel-item van-hairline--surround">{{ obj.name }}
              <van-badge
                content="+"
                color="transparent"
                class="cross-badge"
              >
              </van-badge>
            </div>
          </van-col>
        </van-row>
      </div>
    </div>
</template>
```

（2）在<script>标签中定义传递的 Prop 属性；在 data 选项中定义数据；在 methods 选项中定义方法。其中：edit()方法用于改变 isEdit 的属性值，通过该值控制是否能编辑我的频道列表；more()方法用于向我的频道列表中添加频道，并将该频道从未选择的热门精选频道列表中移除；userChannelClick()方法用于移除我的频道列表中的某个频道，并将该频道添加到未选择的热门精选频道列表中；close()方法用于关闭弹出层。代码如下：

```
<script>
export default {
  props: {
    userList: Array,                          //我的频道列表
    unCheckList: Array,                       //未选择的热门精选频道列表
  },
  data() {
    return {
      isEdit: false,                          //控制编辑状态
    };
  },
  methods: {
    edit() {
```

```
      this.isEdit = !this.isEdit;
    },
    more(channelObj) {
      this.$emit("addChannelEV", channelObj);
    },
    userChannelClick(channelObj) {
      if (this.isEdit === true) {                    //如果处于编辑状态
        if (channelObj.id !== '0') {
          this.$emit("removeChannel", channelObj);
        }
      } else {
        this.$emit("close");                          //关闭弹出层
        this.$emit("toggle", channelObj.id);          //切换到该频道下的新闻列表
      }
    },
    close() {
      this.$emit("close");                            //关闭弹出层
      this.isEdit = false;
    },
  },
};
</script>
```

9.5.5 底部导航栏的设计

底部导航栏主要包括"首页"和"我的"两个选项，其界面效果如图9.14所示。

底部导航栏的主要实现步骤如下。

（1）在 views 文件夹下新建 Layout 文件夹，并在 Layout 文件夹下新建 index.vue 文件。在<template>标签中，定义用于渲染视图的<router-view>组件，同时使用 Tabbar 组件定义底部导航栏，使用 TabbarItem 组件定义导航菜单项。用户单击某个菜单项时，会跳转到相应的页面。代码如下：

图 9.14 底部导航栏的界面效果

```
<template>
  <div>
    <div class="home_container">
      <keep-alive>
        <router-view></router-view>
      </keep-alive>
    </div>
    <div>
      <van-tabbar v-model="active" route>
        <van-tabbar-item name="home" icon="home-o" to="/layout/home">首页</van-tabbar-item>
        <van-tabbar-item name="my" icon="user-o" to="/layout/user">我的</van-tabbar-item
        >
      </van-tabbar>
    </div>
  </div>
</template>
```

（2）在<script>标签中定义 active 属性，该属性用于切换选中的标签，代码如下：

```
<script>
export default {
  name: "LayoutIndex",
  data() {
    return {
      active: 'home'
    };
```

```
    }
  };
</script>
```

9.6 新闻搜索功能的设计与实现

在新闻列表页面，单击右上角的放大镜图标，会进入搜索页面，如图 9.15 所示。在搜索框中输入搜索关键词，然后按 Enter 键或单击搜索框右侧的"搜索"文本，会跳转到搜索结果页面，该页面展示了新闻标题中包含搜索关键词的新闻列表，如图 9.16 所示。此时返回搜索页面，可以看到搜索历史记录，如图 9.17 所示。

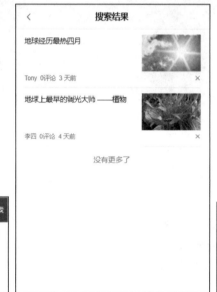
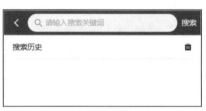

图 9.15　搜索页面　　　　图 9.16　搜索结果页面　　　　图 9.17　显示搜索历史

新闻搜索功能主要涉及两个组件，分别是搜索组件和搜索结果组件，下面对这两个组件进行详细介绍。

9.6.1 搜索组件设计

搜索组件主要用来定义新闻搜索框、"搜索"文本和搜索历史记录列表。搜索组件的主要实现步骤如下。

（1）在 views 文件夹下新建 Search 文件夹，在 Search 文件夹下新建 index.vue 文件。在<template>标签中，使用 Vant 组件分别定义左箭头图标和新闻搜索框。当按 Enter 键时，会调用 search()方法，在 action 插槽中添加"搜索"文本，单击该文本同样会调用 search()方法。然后定义用于删除搜索历史记录的图标，在<span>标签中遍历搜索历史记录数组 history，单击某一项记录会调用 historyClick()方法。代码如下：

```
<template>
  <div>
    <div class="search-header">
      <van-icon
        name="arrow-left"
        color="white"
        size="0.5rem"
        class="go-back"
        @click="$router.back()"
      ></van-icon>
```

```
    <van-search
      v-model="keyword"
      placeholder="请输入搜索关键词"
      background="#0069FF"
      shape="round"
      show-action="true"
      @search="search"
    >
      <template #action>
        <div style="color: #FFFFFF;" @click="search">搜索</div>
      </template>
    </van-search>
  </div>
  <div class="search-history">
    <van-cell title="搜索历史">
      <template #right-icon>
        <van-icon
          name="delete"
          class="search-icon"
          @click="clear"
        ></van-icon>
      </template>
    </van-cell>
    <div class="history-list">
      <span
        class="history-item"
        v-for="(item, index) in history"
        :key="index"
        @click="historyClick(item)"
      >{{ item }}</span>
    </div>
  </div>
</div>
</template>
```

（2）在<script>标签中定义数据，在 methods 选项中定义方法。其中：toSearchResult()方法用于跳转到搜索结果页面；search()方法用于将搜索关键词保存到搜索历史记录数组中，并跳转到搜索结果页面；historyClick()方法用于跳转到搜索结果页面；clear()方法用于清空搜索历史记录。最后在 watch 选项中对搜索历史记录数组进行监听。代码如下：

```
<script>
export default {
  name: "SearchIndex",
  data() {
    return {
      keyword: "",                                              //搜索关键词
      history: JSON.parse(localStorage.getItem("hist")) || []   //搜索历史列表
    };
  },
  methods: {
    toSearchResult(theKeyword) {                                //跳转到搜索结果页面
      this.$router.push({
        path: '/search_result',
        query: { keyword: theKeyword }
      });
    },
    search() {                                                  //按 Enter 键或单击"搜索"文本后触发
      if (this.keyword.length > 0) {
        //将搜索关键字保存到搜索历史记录数组
        this.history.push(this.keyword);
        this.toSearchResult(this.keyword);                      //跳转到搜索结果页面
      }
```

```
        },
        historyClick(str) {
            this.toSearchResult(str);                              //跳转到搜索结果页面
        },
        clear() {                                                  //清空搜索历史记录
            this.history = [];
        },
    },
    watch: {
        history: {                                                 //监听搜索历史记录数组
            deep: true,
            handler() {
                const theSet = new Set(this.history);
                const arr = Array.from(theSet);                    //将集合转换为数组
                localStorage.setItem("hist", JSON.stringify(arr)); //保存搜索历史记录数组
            },
        },
    },
};
</script>
```

9.6.2 搜索结果组件设计

搜索结果组件主要用来遍历搜索到的新闻列表，在该组件中同样使用了上拉加载功能。搜索结果组件的主要实现步骤如下。

（1）在 Search 文件夹下新建 SearchResult.vue 文件。在<template>标签中使用 List 组件实现上拉加载功能，在 List 组件中调用新闻列表项组件 NewsItem.vue，在调用组件时对搜索到的新闻列表数组进行遍历，并将遍历时的每个新闻对象作为 Prop 属性的值传递给 NewsItem.vue 组件。当单击每条新闻时，会调用 itemClick()方法。代码如下：

```
<template>
  <div>
    <div class="search-result-container">
      <van-nav-bar
        title="搜索结果"
        left-arrow
        @click-left="$router.go(-1)"
        fixed
      />
    </div>
    <div>
      <van-list
        v-model:loading="loading"
        :finished="finished"
        :immediate-check="false"
        finished-text="没有更多了"
        @load="onLoad"
      >
        <NewsItem
          v-for="obj in list"
          :key="obj.id"
          :newsObj="obj"
          :isShow="false"
          @click="itemClick(obj.id)"
        ></NewsItem>
      </van-list>
    </div>
  </div>
</template>
```

（2）在<script>标签中引入 axios 实例和新闻列表项组件。在 data 选项中定义数据；在 components 选项中注册新闻列表项组件；在 created()钩子函数中查询 JSON 服务器中的 NewsList 数据接口，获取所有新闻列表，并过滤出新闻标题中包含搜索关键词的新闻列表；在 methods 选项中定义方法。其中：onLoad()方法用于上拉加载数据；itemClick()方法用于跳转到新闻详情页面，并在该方法中将新闻 id 作为参数进行传递。代码如下：

```vue
<script>
import api from "@/utils/request";                         //引入 axios 实例
import NewsItem from "@/components/NewsItem.vue";          //引入新闻列表项组件
export default {
  name: "SearchResult",
  data() {
    return {
      now: 0,                                              //第几页
      limit: 10,                                           //每页显示的数据个数
      totalList: [],                                       //搜索到的整个新闻列表
      list: [],                                            //分页加载时的新闻列表
      loading: false,                                      //是否处于加载状态
      finished: false,                                     //是否已加载完成
      totalNum: 0                                          //搜索到的新闻数量
    };
  },
  components: {
    NewsItem,
  },
  async created() {
    //查询新闻列表
    await api({
      url: '/NewsList',
      method: 'GET'
    }).then((res) => {
      //过滤出新闻标题中包含搜索关键词的新闻列表
      this.totalList = res.data.filter((item) => {
        return item.title.includes(this.$route.query.keyword)
      })
      this.totalNum = this.totalList.length;              //获取搜索到的新闻数量
      this.onLoad();                                      //分页加载
    });
  },
  methods: {
    async onLoad() {
      this.now += 1;                                      //将当前页数加 1
      //加载后的新闻列表
      this.list = this.totalList.slice(0, this.now * this.limit);
      this.loading = false;
      //数据全部加载完成
      if (this.list.length >= this.totalNum) {
        this.finished = true;
      }
    },
    itemClick(id) {                                       //跳转到新闻详情页面
      this.$router.push({
        path: '/detail',
        query: {
          news_id: id                                     //查询参数为新闻 id
        }
      });
    },
  },
};
</script>
```

9.7 新闻详情页面的设计与实现

在新闻列表页面,单击某一条新闻会跳转到对应的新闻详情页面。新闻详情页面展示了这条新闻的详细信息,包括新闻标题、新闻发布用户头像、新闻发布用户名、新闻发布时间到当前时间的时间间隔、新闻内容和用户评论内容等信息,其页面效果如图 9.18 所示。

新闻详情页面主要涉及两个组件,分别是新闻内容组件和用户评论组件。下面对这两个组件分别进行详细介绍。

9.7.1 新闻内容组件设计

新闻内容组件是用户评论组件的父组件,该组件主要定义关于新闻的详细信息,主要包括新闻标题、新闻发布用户头像、新闻发布用户名、新闻发布时间到当前时间的时间间隔和新闻内容,具体实现步骤如下。

图 9.18　新闻详情页面

(1)在 views 文件夹下新建 NewsDetail 文件夹,在 NewsDetail 文件夹下新建 index.vue 文件。在<template>标签中使用 Vant 中的 NavBar 组件定义页面标题,使用 Loading 组件定义加载中的过渡状态。然后分别定义新闻标题、新闻发布用户名、新闻发布时间到当前时间的时间间隔、新闻发布用户头像、"关注"按钮和新闻内容。接下来使用 Popup 组件定义弹出层,当 show 属性值为 true 时显示该弹出层,在弹出层中显示相应的提示信息。最后调用用户评论组件 CommentList.vue,将该新闻对象作为 Prop 属性的值传递给 CommentList.vue 组件。关键代码如下:

```
<template>
  <div>
    <van-nav-bar
      fixed
      title="新闻详情"
      left-arrow
      @click-left="$router.go(-1)"
    />
    <van-loading color="#1567FC" v-if="Object.keys(newsObj).length === 0">
      新闻加载中...
    </van-loading>
    <div v-else>
      <div class="News-container">
        <h1 class="news-title">{{ newsObj.title }}</h1>
        <van-cell
          center
          :title="newsObj.aut_name"
          :label="timeInterval(newsObj.pubDate)"
        >
          <template #icon>
            <img :src="newsObj.aut_photo" alt="" class="avatar" />
          </template>
          <template #default>
```

```html
            <div>
              <van-button style="width: 40px;color: #666666;"
                  size="mini"
                  v-if="isFollowed"
                  @click="toFollow"
                  >已关注</van-button
              >
              <van-button  style="width: 40px;"
                  type="danger"
                  size="mini"
                  plain
                  v-else
                  @click="toFollow"
                  >关注</van-button
              >
            </div>
          </template>
        </van-cell>
        <van-divider></van-divider>
        <div class="news-content" v-html="newsObj.content"></div>
        <van-divider>End</van-divider>
        <div class="like-box"></div>
    </div>
    <van-popup v-model:show="show"
            closeable
            round
            teleport="body"
            style="width: 60%; height: 30%; text-align: center"
    >
      <van-cell>
        <view class="cell-text">{{ message }}</view>
      </van-cell>
      <van-button style="width: 80%;" type="danger" to="/login">一键登录</van-button>
    </van-popup>
    <div>
      <CommentList :newsObj="newsObj"></CommentList>
    </div>
  </div>
 </div>
</template>
```

（2）在<script>标签中分别引入 axios 实例、timeInterval()函数和用户评论组件。在 data 选项中定义数据；在 created()钩子函数中获取登录用户 id，然后分别根据新闻 id 查询新闻详情、根据用户 id 查询关注列表，再对关注列表进行遍历并判断，如果关注列表中的新闻发布用户 id 和当前新闻发布用户 id 相同，就将"关注"按钮显示为"已关注"；在 methods 选项中注册 timeInterval()函数和定义 toFollow()方法，该方法用于关注该新闻发布用户或取消关注；在 components 选项中注册用户评论组件 CommentList.vue。代码如下：

```javascript
<script>
import api from "@/utils/request.js";                          //引入 axios 实例
import { timeInterval } from "@/utils/date.js";                //引入时间间隔函数
import CommentList from "./CommentList.vue";                   //引入评论列表组件
export default {
  name: "NewsDetail",
  data() {
    return {
      newsObj: {},                                             //新闻对象
      isFollowed: false,                                       //是否已关注
      userId: "",                                              //用户 id
      news_id: this.$route.query.news_id,                      //新闻 id
      show: false                                              //是否显示弹出层
    };
```

```js
    },
    async created() {
        //获取用户id
        this.userId = localStorage.getItem("userId") ? localStorage.getItem("userId") : "";
        //根据新闻id查询新闻详情
        const res = await api({
            url: "/NewsDetail",
            method: "GET",
            params: {
                id: this.news_id
            }
        })
        this.newsObj = res.data[0];                                         //获取新闻对象
        //根据用户id查询关注列表
        const res1 = await api({
            url: "/FollowList",
            method: "GET",
            params: {
                userId: this.userId
            }
        });
        //如果关注列表中的新闻发布用户id和该新闻发布用户id相同，就显示为已关注
        res1.data.forEach((item) => {
            if(item.aut_id === this.newsObj.aut_id){
                this.isFollowed = true;
            }
        })
    },
    methods: {
        timeInterval,                                                       //时间间隔函数
        async toFollow(){                                                   //关注或取消关注
            if(this.userId){
                const res = await api({
                    url: `/UserList`,
                    method: "GET",
                    params: {
                        id: this.userId
                    }
                });
                let follow_count = res.data[0].follow_count;                //获取用户信息中的关注数量
                if(!this.isFollowed){
                    //将登录用户id和新闻发布用户id添加到关注列表中
                    await api({
                        url: "/FollowList",
                        method: "POST",
                        data: {
                            userId: this.userId,                            //登录用户id
                            aut_id: this.newsObj.aut_id                     //新闻发布用户id
                        }
                    });
                    this.isFollowed = true;                                 //已关注
                    //更新用户信息中的关注数量加1
                    await api({
                        url: `/UserList/${this.userId}`,
                        method: "PATCH",
                        data: {
                            follow_count: follow_count += 1
                        }
                    });
                }else{
                    //根据登录用户id和新闻发布用户id查询关注列表
                    const res = await api({
```

```
              url: "/FollowList",
              method: "GET",
              params: {
                userId: this.userId,
                aut_id: this.newsObj.aut_id            //新闻发布用户 id
              }
            });
            //将关注记录从关注列表中移除
            await api({
              url: `/FollowList/${res.data[0].id}`,
              method: "DELETE"
            });
            this.isFollowed = false;                    //取消关注
            //更新用户信息中的关注数量减 1
            await api({
              url: `/UserList/${this.userId}`,
              method: "PATCH",
              data: {
                follow_count: follow_count -= 1
              }
            });
          }
        }else{
          this.show = true;
          this.message = "登录后可关注";
        }
      }
    },
    components: {
      CommentList,
    },
  };
</script>
```

> **说明**
> 只有登录用户才能对新闻发布用户进行关注或取消关注。

9.7.2 用户评论组件的设计

用户评论组件是新闻内容组件的子组件，具备遍历用户评论列表、点赞评论、发布评论、点赞新闻和收藏新闻等功能。用户评论界面的效果如图9.19所示。

用户评论组件的主要实现步骤如下。

（1）在 NewsDetail 文件夹下新建 CommentList.vue 文件。在 <template>标签中使用 List 组件实现上拉加载功能。在 List 组件中对用户评论数组 commentArr 进行遍历，在遍历时分别输出发布用户头像、用户名、心形点赞图标、评论内容和评论发布时间到当

图 9.19 用户评论界面效果

前时间的时间间隔。在用户评论列表下方添加"发布评论"的文本信息，单击该文本调用 toggleShow()方法，然后分别添加表示分享、评论、点赞和收藏的 4 个图标和对应的文字。其中，分享图标并没有实际的功能，单击评论图标会调用 commentClick()方法，单击点赞图标会调用 toLikeNews()方法，单击收藏图标会调用 toCollect()方法。在这 4 个图标的下方添加一个用于输入评论内容的文本域和一个"发布"按钮，单击"发布"按钮会调用 send()方法。最后使用 Popup 组件定义弹出层，当 show 属性值为 true 时显示该弹出层，在

弹出层中显示相应的提示信息。代码如下：

```html
<template>
  <div>
    <div
      class="comment-list"
      :class="{
        'set-padding-1': isShowBox,
        'set-padding-2': !isShowBox,
      }"
    >
      <div class="comments-top" v-if="totalCount">评论 {{totalCount}}</div>
      <div class="comments-top" v-else>暂无评论</div>
      <van-list
        v-model:loading="loading"
        :finished="finished"
        :finished-text="commentArr.length ? '已显示全部评论' : '暂无评论，点击抢沙发'"
        @load="onLoad"
      >
        <div class="comment-item" v-for="obj in commentArr" :key="obj.id">
          <div class="comment-header">
            <div class="comment-header-left">
              <img :src="obj.photo" alt="" class="avatar" />
              <span class="uname">{{ obj.username }}</span>
            </div>
            <div class="comment-header-right">
              <view>
                <van-icon v-if="obj.isLikedComment" name="like" size="16" color="red" @click="toLikedComment(obj)"/>
                <van-icon v-else name="like-o" size="16" color="gray" @click="toLikedComment(obj)"/>
                <text>{{obj.like_count ? obj.like_count : "赞"}}</text>
              </view>
            </div>
          </div>
          <div class="comment-body">
            {{ obj.content }}
          </div>
          <div class="comment-footer">{{ timeInterval(obj.pubDate) }}</div>
        </div>
      </van-list>
    </div>
    <div>
      <div class="add-comment-box van-hairline--top" v-if="isShowBox === true">
        <div class="ipt-comment-div" @click="toggleShow">发表评论</div>
        <div class="icon-box">
          <view>
            <van-icon name="share-o" size="0.56rem" /><text>分享</text>
          </view>
          <view>
            <van-icon
              name="comment-o"
              size="0.56rem"
              @click="commentClick"
            /><text>{{totalCount?totalCount:"评论"}}</text>
          </view>
          <view>
            <van-icon v-if="isLikedNews" name="like" size="0.56rem" color="red" @click="toLikeNews"/>
            <van-icon v-else name="like-o" size="0.56rem" @click="toLikeNews"/>
            <text>{{ like_count ? like_count : "点赞" }}</text>
          </view>
          <view>
            <van-icon v-if="isCollect" name="star" size="0.56rem" color="red" @click="toCollect"/>
            <van-icon v-else name="star-o" size="0.56rem" @click="toCollect"/>
```

```html
        <text>{{ collected_count ? collected_count : "收藏" }}</text>
      </view>
    </div>
  </div>
  <div class="comment-box van-hairline--top" v-else>
    <textarea
      placeholder="请输入评论内容"
      @blur="blurFun"
      v-model.trim="commentText"
    ></textarea>
    <van-button style="color: #FF0000;"
      type="default"
      :disabled="commentText.length === 0"
      @click="send"
    >发布</van-button>
  </div>
</div>
<van-popup v-model:show="show"
  closeable
  round
  teleport="body"
  style="width: 60%; height: 30%; text-align: center"
>
  <van-cell>
    <view class="cell-text">{{ message }}</view>
  </van-cell>
  <van-button style="width: 80%;" type="danger" to="/login">一键登录</van-button>
</van-popup>
</div>
</template>
```

（2）在<script>标签中分别引入 axios 实例和 timeInterval()函数。在 props 选项中定义传递的 Prop 属性。在 data 选项中定义数据。在 created()钩子函数中判断用户是否已登录，如果已登录，就查询 JSON 服务器中的 UserList 数据接口，获取当前登录用户信息。接着查询 JSON 服务器中的 NewsLikeList 数据接口，根据结果判断用户是否已点赞该新闻，如果已点赞，就点亮心形点赞图标。之后查询 JSON 服务器中的 CollectList 数据接口，根据结果判断用户是否已收藏该新闻。如果已收藏，就点亮星形收藏图标；如果用户未登录，就不点亮点赞和收藏图标。代码如下：

```javascript
<script>
import api from "@/utils/request";                              //引入 axios 实例
import { timeInterval } from "@/utils/date.js";                 //引入时间间隔函数
export default {
  props: [
    "newsObj"
  ],
  data() {
    return {
      news_id: this.$route.query.news_id,                       //新闻 id
      isLikedNews: false,                                        //是否已点赞该新闻
      isCollect: false,                                          //是否已收藏该新闻
      commentArr: [],                                            //评论列表
      totalCount: 0,                                             //评论总数量
      isShowBox: true,                                           //控制显示哪个评论容器
      commentText: "",                                           //评论内容
      loading: false,                                            //是否处于加载状态
      finished: false,                                           //是否已加载完成
      user: null,                                                //登录用户对象
      show: false,                                               //是否显示弹出层
      message: "",                                               //弹出层中的提示文字
```

```
          like_count: this.newsObj.like_count,           //新闻点赞数量
          collected_count: this.newsObj.collected_count, //新闻收藏数量
          now: 0,                                        //第几页
          limit: 10                                      //每页显示的数据个数
      };
    },
    async created() {
      if(localStorage.getItem("userId")){
        //如果用户已登录，就获取用户信息
        const res1 = await api({
          url: "/UserList",
          method: "GET",
          params: {
            id: localStorage.getItem("userId")
          }
        })
        this.user = res1.data[0];
        //如果用户已登录，就判断是否已点赞该新闻
        const res3 = await api({
          url: "/NewsLikeList",
          method: "GET",
          params: {
            userId: this.user.id,
            news_id: this.news_id
          }
        })
        this.isLikedNews = res3.data.length > 0 ? true : false;
        //如果用户已登录，就判断是否已收藏该新闻
        const res4 = await api({
          url: "/CollectList",
          method: "GET",
          params: {
            userId: this.user.id,
            news_id: this.news_id
          }
        })
        this.isCollect = res4.data.length > 0 ? true : false;
      } else {
        //如果用户未登录，就不点亮点赞和收藏图标
        this.isLikedNews = false;
        this.isCollect = false;
      }
    },
  };
</script>
```

（3）在 created()钩子函数下方定义 methods 选项，在该选项中首先注册 timeInterval()函数，然后定义 initLikeComment()方法，该方法用于初始化用户点赞评论的记录。在 initLikeComment()方法中，查询 JSON 服务器中的 CommentLikeList 数据接口，根据登录用户 id 和新闻 id 查询评论点赞列表，之后通过遍历用户评论列表和评论点赞列表来点亮对应的评论心形点赞图标。代码如下：

```
methods: {
  timeInterval,                                  //时间间隔函数
  async initLikeComment(){                       //初始化用户点赞评论的记录
    const res = await api({                      //根据用户 id 和新闻 id 查询评论点赞列表
      url: "/CommentLikeList",
      method: "GET",
      params: {
        userId: this.user.id,
        news_id: this.news_id
      }
```

```
  })
  this.commentArr.forEach((item) => {                    //遍历评论列表
    item.isLikedComment = false;
    res.data.forEach((item1) => {                        //遍历评论点赞列表
      //如果评论 id 和评论点赞列表的 id 相同,就点亮该评论点赞图标
      if(item.id === item1.com_id){
        item.isLikedComment = true;
      }
    })
  })
},
```

(4) 在 initLikeComment()方法下方定义 toLikedComment()方法,该方法用于点赞用户评论。在 toLikedComment()方法中,判断用户是否已经点赞了该评论。如果用户还未点赞该评论,就将点赞记录添加到评论点赞列表中,同时更新评论列表中该评论的点赞数量,使该数量加 1,之后调用 initLikeComment()方法重新初始化用户点赞评论的记录,实现点亮心形点赞图标的效果;如果用户已点赞该评论,就根据用户 id 和评论 id 查询该评论,再根据评论 id 将点赞记录从评论点赞列表中移除,同时更新评论列表中该评论的点赞数量,使该数量减 1,接着重新初始化用户点赞评论的记录,实现取消点亮心形点赞图标的效果。代码如下:

```
async toLikedComment(obj) {                              //点赞评论
  if(localStorage.getItem("userId")){
    if (!obj.isLikedComment) {                           //如果用户还未点赞该新闻
      //将记录添加到评论点赞列表中
      await api({
        url: "/CommentLikeList",
        method: "POST",
        data: {
          userId: this.user.id,                          //用户 id
          com_id: obj.id,                                //评论 id
          news_id: this.news_id                          //新闻 id
        }
      });
      let count = obj.like_count;                        //获取该评论点赞数量
      //更新评论列表中该评论的点赞数量,使该数量加 1
      await api({
        url: `/CommentsList/${obj.id}`,
        method: "PATCH",
        data: {
          like_count: count += 1
        }
      })
      await this.initLikeComment();                      //重新初始化用户点赞评论的记录,更新点赞图标状态
      this.commentArr.forEach((item) => {                //遍历评论列表
        if(item.id === obj.id){
          item.like_count += 1;                          //更新点赞数量
        }
      })
    } else {
      //根据用户 id 和评论 id 查询该评论
      const res = await api({
        url: "/CommentLikeList",
        method: "GET",
        params: {
          userId: this.user.id,
          com_id: obj.id
        }
      });
      //将点赞记录从评论点赞列表中移除
      await api({
```

```
            url: `/CommentLikeList/${res.data[0].id}`,
            method: "DELETE"
          });
          let count = obj.like_count;                      //获取该评论点赞数量
          //更新评论列表中该评论的点赞数量，使该数量减1
          await api({
            url: `/CommentsList/${obj.id}`,
            method: "PATCH",
            data: {
              like_count: count -= 1
            }
          })
          await this.initLikeComment();                    //重新初始化用户点赞评论的记录，更新点赞图标状态
          this.commentArr.forEach((item) => {              //遍历评论列表
            if(item.id === obj.id){
              item.like_count -= 1;                        //更新点赞数量
            }
          })
        }
      }else{
        this.show = true;
        this.message = "登录后可点赞";
      }
    },
```

（5）在 toLikedComment()方法下方定义 toLikeNews()方法，该方法用于点赞新闻。在 toLikeNews()方法中判断用户是否已经点赞了该新闻。如果用户还未点赞该新闻，就点亮新闻点赞图标，再将点赞记录添加到新闻点赞列表中，同时更新新闻点赞的数量，使该数量加 1；如果用户已点赞该新闻，就取消点亮新闻点赞图标，再根据用户 id 和新闻 id 查询新闻点赞列表，根据 id 将点赞记录从新闻点赞列表中移除，同时更新新闻点赞的数量，使该数量减 1。代码如下：

```
async toLikeNews() {                                       //点赞新闻
    if(localStorage.getItem("userId")){
      if (!this.isLikedNews) {                             //如果用户还未点赞该新闻
        this.isLikedNews = true;                           //点亮新闻点赞图标
        //将记录添加到新闻点赞列表中
        await api({
          url: "/NewsLikeList",
          method: "POST",
          data: {
            userId: this.user.id,
            news_id: this.news_id
          }
        });
        this.like_count += 1;                              //将该新闻点赞数量加1
        //更新新闻点赞数量
        await api({
          url: `/NewsDetail/${this.newsObj.id}`,
          method: "PATCH",
          data: {
            like_count: this.like_count
          }
        })
      } else {
        this.isLikedNews = false;                          //取消点亮新闻点赞图标
        //根据用户id和新闻id查询新闻点赞列表
        const res = await api({
          url: "/NewsLikeList",
          method: "GET",
          params: {
            userId: this.user.id,
```

```js
          news_id: this.news_id
        }
      });
      //将点赞记录从新闻点赞列表中移除
      await api({
        url: `/NewsLikeList/${res.data[0].id}`,
        method: "DELETE"
      });
      this.like_count -= 1;                         //将该新闻点赞数量减1
      //更新新闻点赞数量
      await api({
        url: `/NewsDetail/${this.newsObj.id}`,
        method: "PATCH",
        data: {
          like_count: this.like_count
        }
      })
    }
  }else{
    this.show = true;
    this.message = "登录后可点赞";
  }
},
```

（6）在toLikeNews()方法下方定义toCollect()方法，该方法用于收藏新闻。在toCollect()方法中判断用户是否已经收藏了该新闻。如果用户还未收藏该新闻，就点亮新闻收藏图标，再将收藏记录添加到新闻收藏列表中，同时更新新闻收藏的数量，使该数量加1；如果用户已收藏该新闻，就取消点亮新闻收藏图标，再根据用户id和新闻id查询新闻收藏列表，根据id将收藏记录从新闻收藏列表中移除，同时更新新闻收藏的数量，使该数量减1。代码如下：

```js
async toCollect() {                                 //收藏新闻
    if(localStorage.getItem("userId")){
      if (!this.isCollect) {                        //如果用户还未收藏该新闻
        this.isCollect = true;                      //点亮新闻收藏图标
        //将记录添加到新闻收藏列表中
        await api({
          url: "/CollectList",
          method: "POST",
          data: {
            userId: this.user.id,
            news_id: this.news_id
          }
        });
        this.collected_count += 1;                  //将该新闻收藏数量加1
        //更新新闻收藏数量
        await api({
          url: `/NewsDetail/${this.newsObj.id}`,
          method: "PATCH",
          data: {
            collected_count: this.collected_count
          }
        })
      } else {
        this.isCollect = false;                     //取消点亮新闻收藏图标
        //根据用户id和新闻id查询新闻收藏列表
        const res = await api({
          url: "/CollectList",
          method: "GET",
          params: {
            userId: this.user.id,
            news_id: this.news_id
```

```js
            }
        });
        //将收藏记录从新闻收藏列表中移除
        await api({
            url: `/CollectList/${res.data[0].id}`,
            method: "DELETE"
        });
        this.collected_count -= 1;              //将该新闻收藏数量减1
        //更新新闻收藏数量
        await api({
            url: `/NewsDetail/${this.newsObj.id}`,
            method: "PATCH",
            data: {
                collected_count: this.collected_count
            }
        })
    }
}else{
    this.show = true;
    this.message = "登录后可收藏";
}
},
```

（7）在 toCollect()方法下方定义 toggleShow()方法，该方法用于控制评论发布区域的显示状态。在 toggleShow()方法下方定义 commentClick()方法，该方法用于将评论列表滑动到第一条评论处。在 commentClick()方法下方定义 send()方法，该方法用于发布评论，发布评论后，该方法会将评论信息添加到评论列表数组中，从而实现更新视图的效果。在 send()方法下方定义 blurFun()方法，该方法用于隐藏发布评论的文本域和"发布"按钮。在 blurFun()方法下方定义 onLoad()方法，该方法用于加载评论列表，在显示评论列表时以发布时间降序排列。代码如下：

```js
//控制评论发布区域的显示状态
toggleShow() {
    if(!localStorage.getItem("userId")){
        this.show = true;
        this.message = "登录后可评论";
    }else{
        this.isShowBox = false;
    }
},
//滑动到第一条评论处
commentClick() {
    document.querySelector(".like-box").scrollIntoView({
        behavior: "smooth"
    });
},
async send() {                                  //发布评论
    //将评论信息添加到评论列表中
    const res1 = await api({
        url: "/CommentsList",
        method: "POST",
        data: {
            news_id: this.news_id,
            username: this.user.username,
            photo: this.user.photo,
            like_count: 0,
            content: this.commentText,
            pubDate: new Date().toLocaleString()
        }
    })
    //将评论信息添加到评论列表数组中，以更新视图
```

```js
            this.commentArr.unshift(res1.data);
            this.totalCount++;
            this.commentText = "";
            this.commentClick();                        //滑动到第一条评论处
            //根据新闻 id 查询新闻信息
            const res2 = await api({
                url: "/NewsList",
                method: "GET",
                params: {
                    id: this.news_id
                }
            })
            //根据新闻 id 更新新闻评论数量
            await api({
                url: `/NewsList/${this.news_id}`,
                method: "PATCH",
                data: {
                    comm_count: res2.data[0].comm_count + 1
                }
            })
        },
        blurFun() {
            setTimeout(() => {
                this.isShowBox = true;
            }, 10);
        },
        async onLoad() {                                //加载数据
            const res = await api({
                url: "/CommentsList",
                method: "GET",
                params: {
                    news_id: this.news_id               //新闻 id
                }
            });
            this.totalCount = res.data.length;          //评论总数
            if (this.totalCount === 0) {
                this.finished = true;
            } else {
                this.now += 1;
                await api({
                    url: '/CommentsList?_sort=-pubDate',  //以发布时间降序排列
                    method: 'GET',
                    params: {
                        news_id: this.news_id,
                        _page: this.now,
                        _per_page: this.limit
                    }
                }).then((res) => {
                    this.loading = false;
                    this.commentArr = [...this.commentArr, ...res.data.data];
                    //finished 为 true，页面到底后不再触发 load 事件
                    if (this.commentArr.length >= this.totalCount) {
                        this.finished = true;
                    }
                });
            }
            //如果用户已登录，就初始化用户点赞评论的记录
            if(localStorage.getItem("userId")){
                await this.initLikeComment();
            }
        }
```

> **说明**
> 只有登录用户才能进行点赞新闻、收藏新闻和点赞评论的操作。

9.8 注册和登录页面的设计与实现

在仿今日头条 APP 中，用户只有在成功登录之后，才能进行点赞新闻或点赞用户评论等操作。在登录之前，用户如果尚未注册，就需要先进行注册。用户注册页面和登录页面的效果分别如图 9.20 和图 9.21 所示。

图 9.20 用户注册页面效果

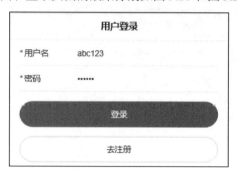

图 9.21 用户登录页面效果

9.8.1 注册页面的设计

在用户注册页面中，用户需要分别输入用户名、密码和确认密码。如果输入无误，单击"注册"按钮即可完成注册。注册页面的主要实现步骤如下。

（1）在 views 文件夹下新建 Register 文件夹，并在 Register 文件夹下新建 index.vue 文件。在<template>标签中，使用 Vant 中的 Form 组件和 Field 组件定义表单和表单元素，在表单元素中，通过 rules 属性定义表单校验规则。接着，使用 Button 组件定义一个"注册"按钮，当单击"注册"按钮时，会触发表单的 submit 事件，同时调用 onSubmit()方法进行注册。代码如下：

```
<template>
  <div>
    <van-nav-bar title="用户注册" left-arrow @click-left="$router.go(-1)" />
    <van-form @submit="onSubmit">
      <van-field
        required
        v-model="user.username"
        name="username"
        label="用户名"
        placeholder="用户名"
        :rules="[
          {
            required: true,
            message: '用户名只能包括中文、字母或数字',
            pattern: /^[\u4e00-\u9fa5a-zA-Z0-9]+$/,
          },
        ]"
      />
      <van-field
        required
```

```
            v-model="user.pwd"
            type="password"
            name="pwd"
            label="密码"
            placeholder="密码"
            :rules="[
              { required: true, message: '密码由数字或字母组成，且至少 6 位', pattern: /^[0-9A-Za-z]{6,}$/ },
            ]"
          />
          <van-field
              required
              v-model="confirmPwd"
              type="password"
              name="confirmPwd"
              label="确认密码"
              placeholder="确认密码"
              :rules="[{  required: true, validator: validatorMessage, message: '请输入确认密码' }]"
          />
          <div style="margin: 16px">
            <van-button
              round
              block
              type="primary"
              native-type="submit"
              :disabled="isloading"
              :loading="isloading"
              loading-text="加载中..."
              >注册</van-button
            >
          </div>
        </van-form>
      </div>
</template>
```

（2）在<script>标签中分别引入 axios 实例和 Vant 中的消息通知组件，然后在 data 选项中定义数据，并在 methods 选项中定义方法。其中：validatorMessage()方法用于验证两次输入的密码是否一致；onSubmit()方法首先判断输入的用户名是否已经存在，如果已存在，就给出提示，否则就随机为注册用户分配一个头像，再将用户信息添加到用户列表中，之后跳转到用户登录页面。代码如下：

```
<script>
import api from "@/utils/request";                      //引入 axios 实例
import { showNotify } from "vant";                      //引入 Vant 中的消息通知组件
export default {
  name: "LoginIndex",
  data() {
    return {
      user: {
        username: "",                                   //用户名
        pwd: "",                                        //密码
        photo: ""                                       //头像
      },
      confirmPwd: '',                                   //确认密码
      isloading: false                                  //控制按钮是否可用和按钮的加载状态
    };
  },
  methods: {
    validatorMessage(){                                 //验证两次输入的密码是否一致
      if(this.confirmPwd && this.user.pwd !== this.confirmPwd){
        return "两次密码不一致";
      }
    },
```

```
    async onSubmit(){
      this.isloading = true;
      //根据用户名查找用户
      await api.get(`/UserList?username=${this.user.username}`).then((response) => {
        if (response.data.length > 0) {                    //如果有查询结果
          this.user.username = "";
          showNotify({ type: "danger", message: "该用户名已存在" });
        } else {
          let rand = Math.ceil(Math.random() * 10);        //生成一个 1~10 的随机数
          this.user.photo = "/images/head/" + rand + ".gif"; //根据生成的随机数分配用户头像
          this.user.follow_count = 0;                      //关注数量
          this.user.fans_count = 0;                        //被关注数量
          this.user.like_count = 0;                        //获赞数量
          //将用户信息添加到用户列表中
          api.post("/UserList", this.user).then((response) => {
            if (response.data.id) {
              showNotify({ type: "success", message: "注册成功" });
              setTimeout(() => {
                this.$router.push({ path: "/login" });     //注册后，跳转到用户登录页面
              }, 1000);
            }
          });
        }
      });
      this.isloading = false;
    }
  },
};
</script>
```

9.8.2 登录页面的设计

用户注册成功后会跳转到用户登录页面。在用户登录页面中，用户需要分别输入用户名和密码，如果输入正确，单击"登录"按钮即可登录成功，登录成功后会跳转到新闻列表页面。登录页面的主要实现步骤如下。

（1）在 views 文件夹下新建 Login 文件夹，在 Login 文件夹下新建 index.vue 文件。在<template>标签中使用 Vant 中的 Form 组件和 Field 组件定义表单和表单元素。在表单元素中通过 rules 属性定义表单校验规则。接着，使用 Button 组件分别定义一个"登录"按钮和一个"去注册"按钮。当用户单击"登录"按钮时，会触发表单的 submit 事件，同时调用 onSubmit()方法进行登录；当用户单击"去注册"按钮时，会调用 toReg()方法。代码如下：

```
<template>
  <div>
    <van-nav-bar title="用户登录" />
    <van-form @submit="onSubmit">
      <van-field
        required
        v-model="user.username"
        name="username"
        label="用户名"
        placeholder="用户名"
        :rules="[
          {
            required: true,
            message: '用户名只能包括中文、字母或数字',
            pattern: /^[\u4e00-\u9fa5a-zA-Z0-9]+$/,
          },
        ]"
```

```
        />
        <van-field
          required
          v-model="user.pwd"
          type="password"
          name="pwd"
          label="密码"
          placeholder="密码"
          :rules="[
            { required: true, message: '密码由数字或字母组成，且至少 6 位', pattern: /^[0-9A-Za-z]{6,}$/ },
          ]"
        />
        <div style="margin: 16px">
          <van-button
            round
            block
            type="primary"
            native-type="submit"
            :disabled="isloading"
            :loading="isloading"
            loading-text="加载中..."
          >登录</van-button>
        </div>
        <div style="margin: 16px">
          <van-button
            round
            block
            :disabled="isloading"
            :loading="isloading"
            loading-text="加载中..."
            @click="toReg"
          >去注册</van-button>
        </div>
      </van-form>
    </div>
</template>
```

（2）在<script>标签中分别引入 axios 实例和 Vant 中的消息通知组件。接着，在 data 选项中定义数据，在 methods 选项中定义方法。其中，onSubmit()方法会根据输入的用户名和密码对用户信息列表进行查询。如果有查询结果就提示"登录成功"，再将用户 id 保存在 localStorage 中，之后跳转到新闻列表页面；如果没有查询结果，就提示"用户名或密码错误"。在 onSubmit()方法下方定义 toReg()方法，调用该方法可以跳转到用户注册页面。代码如下：

```
<script>
import api from "@/utils/request";              //引入 axios 实例
import { showNotify } from "vant";              //引入 Vant 中的消息通知组件
export default {
  name: "LoginIndex",
  data() {
    return {
      user: {
        username: "",                            //用户名
        pwd: "",                                 //密码
      },
      isloading: false,                          //控制按钮是否可用和按钮的加载状态
    };
  },
  methods: {
    async onSubmit() {
      this.isloading = true;
      //查询用户信息列表
```

```
      await api({
        url: '/UserList',
        method: 'GET',
        params: {
          username: this.user.username,
          pwd: this.user.pwd
        }
      }).then((res) => {
        if(res.data.length !== 0){
          //提示登录成功
          showNotify({ type: "success", message: "登录成功" });
          localStorage.setItem("userId", res.data[0].id);    //保存用户 id
          this.$router.replace({ path: "/layout/home" });    //跳转到新闻列表页面
        }else{
          //提示错误提示信息
          showNotify({ type: "danger", message: "用户名或密码错误" });
        }
      });
      this.isloading = false;
    },
    toReg(){
      thic.$router.rcplace({ path: "/register" }),           //跳转到用户注册页面
    }
  },
};
</script>
```

9.9 我的页面的设计与实现

在底部导航栏中，用户登录成功后，单击"我的"选项会跳转到我的页面。该页面主要展示当前登录用户的关注数量、粉丝数量、获赞数量，以及一些功能选项和"退出登录"文本。该页面效果如图 9.22 所示。

我的页面的主要实现步骤如下。

（1）在 views 文件夹下新建 User 文件夹，在 User 文件夹下新建 index.vue 文件。在 <template> 标签中使用 Vant 中的 Cell 组件定义用户头像和用户名，在 <div> 标签中定义用户的关注数量、粉丝数量和获赞数量。下面定义一些功能选项和"退出登录"文本，单击该文本调用 quit() 方法。代码如下：

```
<template>
  <div class="user-container">
    <div class="user-card">
      <van-cell>
        <template #icon>
          <img :src="userObj.photo" alt="" class="avatar" />
        </template>
        <template #title>
          <div class="username">{{ userObj.username }}</div>
        </template>
      </van-cell>
      <div class="user-data">
        <div class="user-data-item">
          <span>{{ userObj.follow_count }}</span>
          <span>关注</span>
        </div>
        <div class="user-data-item">
```

图 9.22 我的页面

```html
        <span>{{ userObj.fans_count }}</span>
        <span>粉丝</span>
      </div>
      <div class="user-data-item">
        <span>{{ userObj.like_count }}</span>
        <span>获赞</span>
      </div>
    </div>
  </div>
  <div class="action-card">
    <van-button style="margin-right: 10px;">申请认证</van-button>
    <van-button>编辑资料</van-button>
  </div>
  <div class="icon-text">
    <div>
      <van-icon name="chat-o" size="20" />
      <p>消息私信</p>
    </div>
    <div>
      <van-icon name="clock-o" size="20"/>
      <p>浏览历史</p>
    </div>
    <div>
      <van-icon name="like-o"   size="20"/>
      <p>点赞</p>
    </div>
    <div>
      <van-icon name="arrow-down"   size="20"/>
      <p>下载管理</p>
    </div>
  </div>
  <div class="icon-text" style="margin-top: 20px;">
    <div>
      <van-icon name="idcard"   size="20"/>
      <p>我的订单</p>
    </div>
    <div>
      <van-icon name="records-o"   size="20"/>
      <p>关注</p>
    </div>
    <div>
      <van-icon name="pending-payment"   size="20"/>
      <p>钱包</p>
    </div>
    <div>
      <van-icon name="more-o"   size="20"/>
      <p>全部</p>
    </div>
  </div>
  <van-cell style="margin-top: 20px;" icon="warning-o" title="退出登录" is-link @click="quit" />
</div>
</template>
```

（2）在<script>标签中分别引入 axios 实例和 Vant 中的 showConfirmDialog()函数。接着，在 data 选项中定义数据，在 created()钩子函数中根据登录用户 id 查询用户信息，在 methods 选项中定义 quit()方法，该方法通过调用 showConfirmDialog()函数弹出确认对话框。如果用户单击对话框中的"确认"按钮，系统将使用 localStorage 中的 clear()方法清空本地保存的数据，并随后跳转到用户登录页面。代码如下：

```
<script>
import api from "@/utils/request";                    //引入 axios 实例
import { showConfirmDialog } from "vant";             //引入 Vant 中的 showConfirmDialog()函数
```

```js
export default {
  name: "UserIndex",
  data() {
    return {
      userId: localStorage.getItem("userId"),      //用户 id
      userObj: {}                                   //用户信息对象
    };
  },
  async created() {
    //根据用户 id 查询用户信息
    const res = await api({
      url: "/UserList",
      method: "GET",
      params: {
        id: this.userId
      }
    });
    this.userObj = res.data[0];                     //获取用户信息对象
  },
  methods: {
    quit() {
      chowConfirmDialog ({
        title: "是否退出登录",
        message: "确定退出？"
      }).then(() => {
        localStorage.clear();                       //清空本地保存的数据
        this.$router.replace("/login");             //跳转到用户登录页面
      }).catch(() => {});
    }
  }
};
</script>
```

9.10 路由配置

下面给出该项目的路由配置，包括定义路由、创建路由对象、使用 router.beforeEach()注册全局前置守卫。当用户访问我的页面时，该守卫会进行判断：如果用户未登录，就跳转到登录页面。代码如下：

```js
import { createRouter, createWebHistory } from 'vue-router'
const routes = [
  {
    path: "/",
    redirect: "/layout/home",
  },
  {
    path: "/login",
    component: () => import("../views/Login"),
  },
  {
    path: "/register",
    component: () => import("../views/Register"),
  },
  {
    path: "/search",
    component: () => import("../views/Search"),
  },
  {
```

```
      path: "/search_result",
      component: () => import("../views/Search/SearchResult.vue"),
    },
    {
      path: "/detail",
      component: () => import("../views/NewsDetail"),
    },
    {
      path: "/layout",
      component: () => import("../views/Layout"),
      children: [
        {
          path: "user",
          component: () => import("../views/User"),
        },
        {
          path: "home",
          component: () => import("../views/Home")
        },
      ],
    },
];
const router = createRouter({
  history: createWebHistory(process.env.BASE_URL),
  routes,
});
router.beforeEach(function(to,from,next){
  let userId = localStorage.getItem('userId');
  if(to.path==='/layout/user'){                    //访问我的页面
    if(!userId){                                   //如果用户未登录,就跳转到登录页面
      router.push('/login');
    }
  }
  next();
});
export default router;
```

9.11 项目运行

通过前述步骤,我们已经设计并完成了"仿今日头条 APP"项目的开发。接下来,我们运行该项目,以检验我们的开发成果。首先打开 package.json 文件,在 scripts 配置选项中添加一个脚本命令,添加后的代码如下:

```
"scripts": {
  "serve": "vue-cli-service serve",
  "build": "vue-cli-service build",
  "lint": "vue-cli-service lint",
  "mock": "json-server data.json --port 3000"
}
```

打开命令提示符窗口,切换到项目目录,执行启动 JSON Server 的命令,如图 9.23 所示。

新打开一个命令提示符窗口,切换到项目所在的目录,执行 npm run serve 命令运行该项目,如图 9.24 所示。

在浏览器地址栏中输入 http://localhost:8080,然后按 Enter 键,即可跳转到新闻列表页面,其页面效果如图 9.25 所示。进入新闻详情页面后,用户可以对用户评论进行点赞,也可以对新闻进行点赞和收藏,效果如图 9.26 所示。

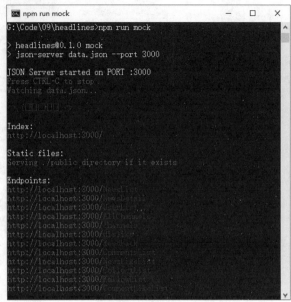

图 9.23　启动 JSON Server

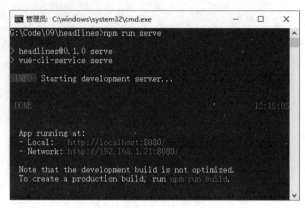

图 9.24　运行项目

图 9.25　仿今日头条 APP 新闻列表页面

图 9.26　点赞和收藏效果

9.12　源码下载

本章虽然详细地讲解了如何编码实现"仿今日头条 APP"的各个功能，但给出的代码都是代码片段，而非完整的源代码。为了方便读者学习，本书提供了该项目的完整源代码，读者可以扫描右侧的二维码进行下载。

源码下载

第 10 章 四季旅游信息网

——Vue CLI + Vue Router + axios + JSON Server + ElementPlus + Day.js

目前，旅游行业十分兴旺，越来越多的人开始把旅游作为主要的休闲娱乐方式。随着互联网的普及，不断有机构开发并运营旅游类网站，这为人们寻找旅游信息和预订旅游产品提供了便利。本章将使用 Vue CLI、Vue Router、axios、JSON Server、ElementPlus 等技术开发一个旅游类网站——四季旅游信息网。

项目微视频

本项目的核心功能及实现技术如下：

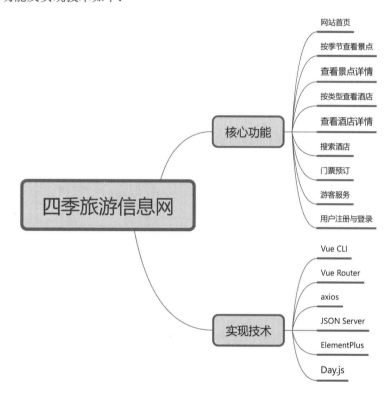

10.1 开发背景

高速发展的互联网，带给人们的不仅仅是技术，更是一种以信息为标志的崭新的生活方式。旅游业是一类对信息技术依赖性很强的产业。信息是旅游业的基础，信息化是旅游业发展的强大推动力。随着经济的发展和人们生活水平的不断提高，旅游活动已经成为人们生活的重要组成部分。旅游者在旅游之前都会收集一些和目的地有关的旅游信息，如当地的景区景点位置、景区活动项目，以及旅游目的地的交通和住宿等情况。因此，为了更好地利用旅游资源，吸引更多的游客前来旅游，开发并运营旅游信息网站，为游客提供全面的旅游信息服务是非常必要的。

本章将使用 Vue CLI、Vue Router、axios、JSON Server、ElementPlus 等技术开发一个旅游信息网站，其实现目标如下：

- ☑ 设计旅游广告轮播图。
- ☑ 按季节查询热门景点。
- ☑ 为各景点设计详情展示。
- ☑ 按酒店类型查询酒店。
- ☑ 为各酒店设计详情展示。
- ☑ 提供酒店搜索功能。
- ☑ 实现门票预订功能。
- ☑ 实现游客服务功能。
- ☑ 实现用户注册和登录。

10.2　系统设计

10.2.1　开发环境

本项目的开发及运行环境如下：

- ☑ 操作系统：推荐 Windows 10、Windows 11 或更高版本，同时兼容 Windows 7（SP1）。
- ☑ 开发工具：WebStorm。
- ☑ 开发框架：Vue.js 3.0。

10.2.2　业务流程

四季旅游信息网由多个页面组成，包括网站首页、热门景点页面、酒店住宿页面、门票预订页面、游客服务页面和用户中心页面等。根据该项目的业务需求，我们设计如图 10.1 所示的业务流程图。

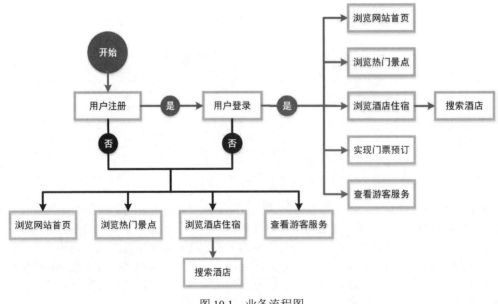

图 10.1　业务流程图

10.2.3 功能结构

本项目的功能结构已经在章首页中给出，其实现的具体功能如下：
- ☑ 网站首页：页面提供网站导航、旅游广告轮播图、热门景点图片展示和酒店住宿图片展示等内容。
- ☑ 热门景点页面：该页面采用分页形式展示景点主图，并按季节对景点进行分类。用户可单击不同的季节选项，查看对应季节的景点。单击任一景点主图，即可进入该景点的详情页面。
- ☑ 景点详情页面：该页面展示各景点的详细信息。
- ☑ 酒店住宿页面：该页面采用分页形式展示酒店信息，并按经济型、舒适型、豪华型等对酒店进行分类。单击不同的类型选项，查看对应类型的酒店。单击酒店名称，即可进入该酒店的详情页面。
- ☑ 酒店详情页面：该页面展示各酒店的详细信息。
- ☑ 酒店搜索结果页面：酒店住宿页面提供一个搜索框，用户可在搜索框中输入搜索关键词，单击"搜索"按钮，进入酒店搜索结果页面，该页面会展示酒店名称中包含搜索关键词的所有酒店信息。
- ☑ 门票预订页面：用户成功登录后，可以进入门票预订页面。在门票预订页面，用户可以选择购买门票的日期、设置购买门票张数。另外，该页面还提供了新增游客信息的功能。
- ☑ 游客服务页面：在游客服务页面，用户可以查看导游信息和游客须知。
- ☑ 用户注册：在注册时，用户需要分别输入用户名、密码、确认密码和邮箱，系统将验证用户输入的内容是否符合要求。
- ☑ 用户登录：在登录时，用户需要分别输入用户名和密码，系统将验证用户输入的用户名和密码是否正确。

10.3 技术准备

10.3.1 技术概览

在开发四季旅游信息网时，我们主要应用 Vue CLI 脚手架工具、Vue Router、axios、JSON Server、ElementPlus 和 Day.js 等技术。Vue CLI 脚手架工具、Vue Router、axios 和 JSON Server 已经在前面的章节做了简要介绍。其中，Vue CLI 脚手架工具、Vue Router 和 axios 在《Vue.js 从入门到精通》中有详细的讲解，对这些知识不太熟悉的读者，可以参考该书对应的内容。Day.js 在第 9 章中已进行了简要的介绍，但该项目主要使用 Day.js 中的另外两个方法——add()和 format()。下面将对 ElementPlus 和 Day.js 中的 add()方法和 format()方法进行必要介绍，以确保读者可以顺利完成本项目。

10.3.2 ElementPlus

1. ElementPlus 简介

ElementPlus 是一套为开发者、设计师和产品经理准备的基于 Vue 3.0 的组件库，它提供了一系列丰富的组件，这些组件旨在帮助开发者快速构建高质量的 Vue 应用程序。作为 ElementUI 的升级版本，ElementPlus 还提供了配套设计资源，助力开发者实现网站的快速成型。

2. 安装 ElementPlus

要在现有的项目中使用 ElementPlus，可以通过 npm 方式进行安装。安装 ElementPlus 的命令如下：

```
npm install element-plus --save
```

3. 使用组件

要使用 ElementPlus 中的组件，需要先引入组件。ElementPlus 组件的引入方式主要有两种，用户可以根据实际业务需要进行选择。

1）全局引入组件

全局引入组件需要在 main.js 文件中进行设置。采用这种方式，我们可以在项目中的任意子组件中使用注册的 ElementPlus 组件。全局引入组件的代码如下：

```
import { createApp } from 'vue'
import App from './App.vue'
import ElementPlus from 'element-plus'
import 'element-plus/dist/index.css'
const app = createApp(App)
app.use(ElementPlus)
app.mount('#app')
```

引入完成后，我们就可以在项目中使用 ElementPlus 提供的任意组件了。

> 说明
>
> index.css 是 ElementPlus 的样式文件。引入该样式文件后，就可以在页面中正常显示 ElementPlus 组件的样式。

2）按需引入组件

如果只想使用 ElementPlus 中的一小部分组件，可以采用按需引入的方式。例如，在组件中引入 ElementPlus 中的 Button 按钮组件，代码如下：

```
import { ElButton } from 'element-plus';
```

按需引入组件后，如果想使用这个组件，还需要在 components 选项中进行注册。例如，注册 Button 按钮组件的代码如下：

```
export default {
    components: { ElButton }
}
```

注册后，我们就可以在模板中使用该组件了。例如，在模板中添加一个"提交"按钮，代码如下：

```
<template>
  <el-button>提交</el-button>
</template>
```

4. ElementPlus 常用组件

ElementPlus 包含了大量的组件。在四季旅游信息网中，我们主要应用了 ElementPlus 中的 Form 组件、FormItem 组件、Input 组件、Button 组件、ElMessage 组件和 DatePicker 组件。下面对这些组件进行介绍。

1）Form 组件

Form 组件用于创建表单。它提供了许多内置的验证规则和验证方法，使表单验证更加容易。使用 Form 组件可以将表单控件组织在一起，并对表单进行验证，以确保提交的数据符合预期的格式和要求。

Form 组件的常用属性及其说明如表 10.1 所示。

表 10.1　Form 组件的常用属性及其说明

属性	说明	类型	默认值
model	用于绑定表单数据对象	object	无默认值
rules	表单验证规则	object	无默认值

续表

属性	说明	类型	默认值
inline	行内表单模式	boolean	false
label-position	表单域标签的位置，当设置为 left 或 right 时，也需要设置 label-width 属性	enum	right
label-width	标签的宽度	string / number	''
disabled	是否禁用该表单内的所有组件。如果设置为 true，将覆盖内部组件的 disabled 属性	boolean	false

Form 组件的常用方法及其说明如表 10.2 所示。

表 10.2 Form 组件的常用方法及其说明

方法	说明
validate	对整个表单进行校验，参数为一个回调函数。该回调函数会在校验结束后被调用，并传入两个参数：是否校验成功和未通过校验的字段。若不传入回调函数，则会返回一个 Promise 对象
validateField	对部分表单字段进行校验
resetFields	对整个表单进行重置，将所有字段值重置为初始值并移除校验结果

2）FormItem 组件

在 Form 组件中，FormItem 组件用于定义表单域，在这些表单域中可以放置各种类型的表单控件，如 Input、Select、Checkbox、Radio、DatePicker、TimePicker 等。

3）Input 组件

Input 是一个可输入的组件，主要用于接收用户的文本输入，它有以下几个作用：

- ☑ 获取用户输入：Input 组件可以用于获取用户在输入框中输入的文本内容。
- ☑ 数据绑定：Input 组件可以与 Vue 实例中的数据进行双向绑定。当用户在输入框中输入文本时，Vue 实例中的数据会自动更新；同样，如果 Vue 实例中的数据发生变化，输入框中的内容也会相应进行更新。
- ☑ 表单验证：Input 组件提供了一些属性和方法，使得表单验证变得简单方便。
- ☑ 输入框样式定制：Input 组件提供了丰富的样式和配置选项，使得输入框的外观和交互效果的定制变得便捷。例如，开发者可以设置输入框的大小、边框样式、占位符文字等。

4）Button 组件

Button 组件用来定义常用的操作按钮。在使用该组件时，开发者可以设置按钮的样式、禁用状态等功能。Button 组件的常用属性及其说明如表 10.3 所示。

表 10.3 Button 组件的常用属性及其说明

属性	说明	类型	默认值
size	按钮的尺寸	enum	无默认值
type	按钮的类型	enum	无默认值
plain	是否为朴素按钮	boolean	false
text	是否为文字按钮	boolean	false
bg	是否显示文字按钮背景颜色	boolean	false
round	是否为圆角按钮	boolean	false
circle	是否为圆形按钮	boolean	false
disabled	按钮是否为禁用状态	boolean	false
color	自定义按钮颜色，并自动计算 hover 和 active 触发后的颜色	string	无默认值

以上 4 个组件都是表单中常用的组件。在下面的例子中，我们将定义一个简单用户登录表单，并设置表单的验证规则。如果用户输入的内容符合要求，系统将提示"登录成功"；否则，系统提示"表单验证失败"。代码如下：

```
<template>
  <el-form :model="form" :rules="rules" ref="form" label-width="100px">
    <el-form-item label="用户名" prop="username">
      <el-input v-model="form.username"></el-input>
    </el-form-item>
    <el-form-item label="密码" prop="password">
      <el-input type="password" v-model="form.password"></el-input>
    </el-form-item>
    <el-form-item>
      <el-button type="primary" @click="submitForm()">登录</el-button>
    </el-form-item>
  </el-form>
</template>
<script>
  export default {
    data() {
      return {
        form: {
          username: '',                                    //用户名
          password: ''                                     //密码
        },
        //定义验证规则
        rules: {
          username: [
            { required: true, message: '请输入用户名', trigger: 'blur' }
          ],
          password: [
            { required: true, message: '请输入密码', trigger: 'blur' },
            { min: 6, max: 12, message: '密码为 6 到 12 个字符', trigger: 'blur' }
          ]
        }
      };
    },
    methods: {
      submitForm(formName) {
        this.$refs.form.validate((valid) => {
          if (valid) {
            alert('登录成功!');
          } else {
            alert('表单验证失败');
            return false;
          }
        });
      }
    }
  }
</script>
```

5）ElMessage 组件

ElMessage 组件用于展示操作反馈信息。该反馈信息从页面顶部出现，3 秒后自动消失。在 Vue 组件中使用 ElMessage 组件时，需要使用 import 语句进行引入，引入的代码如下：

```
import {ElMessage} from "element-plus";
```

ElMessage()方法接收一个对象作为参数来配置消息提示的各种属性，如 message、type、duration 等。其中：message 属性用于定义消息文字；type 属性用于定义消息类型，其可选值有 success、warning、info（默认值）、error；duration 属性用于定义消息的显示时间，单位为毫秒。

例如，定义一个"显示消息"按钮，单击该按钮调用 showMessage()方法，在该方法中显示一个成功类型的消息提示。代码如下：

```
<template>
  <button @click="showMessage">显示消息</button>
</template>
<script>
import { ElMessage } from 'element-plus';          //引入 ElMessage
export default {
  methods: {
    showMessage() {
      ElMessage({
        message: '操作成功',
        type: 'success',
      })
    }
  }
};
</script>
```

6）DatePicker 组件

DatePicker 是日期选择器组件，用于选择或输入日期。日期选择器会在用户未选择任何日期的时候默认展示当天的日期。可以使用 default-value 属性来修改这个默认的日期。

DatePicker 组件的常用属性及其说明如表 10.4 所示。

表 10.4　DatePicker 组件的常用属性及其说明

属　　性	说　　明	类　　型	默　认　值
model-value / v-model	绑定值，如果它是数组，长度应该是 2	number / string / object	''
type	显示类型	enum	date
format	显示在输入框中的格式	string	YYYY-MM-DD
readonly	设置是否只读	boolean	false
disabled	设置是否禁用	boolean	false
default-value	选择器打开时默认显示的时间	object	无默认值
placeholder	非范围选择时的占位内容	string	''

DatePicker 组件的常用事件及其说明如表 10.5 所示。

表 10.5　DatePicker 组件的常用事件及其说明

事　　件	说　　明
change	用户确认选定的值时触发
blur	当 input 失去焦点时触发
focus	当 input 获得焦点时触发

例如，使用 DatePicker 组件定义一个日期选择器，在日期选择器中设置显示日期的格式，代码如下：

```
<el-date-picker
    v-model="date"
    format="M 月 D 日"
    style="width:110px;font-size:12px"
    type="date"
    placeholder="其他日期"
>
</el-date-picker>
```

10.3.3　Day.js 中的 add()方法和 format()方法

1. add()方法

add()方法用于返回一个在原始 Day.js 对象基础上增加了指定时间单位的 Day.js 对象。例如，分别获取 7 天后的 Day.js 对象和 3 小时后的 Day.js 对象，代码如下：

```
dayjs().add(7, 'day');
dayjs().add(3, 'hour');
```

2. format()方法

format()方法用于根据传入的占位符返回格式化后的日期。例如，使用 format()方法输出不同格式的当前日期和时间，代码如下：

```
console.log(now.format('YYYY-MM-DD'));                    //输出当前日期
console.log(now.format('YYYY-MM-DD HH:mm:ss'));           //输出当前日期和时间
console.log(now.format('YYYY 年 M 月 D 日 dddd'));         //输出当前日期和星期
```

10.4　创建项目

在设计四季旅游信息网各功能模块之前，需要使用 Vue CLI 创建项目，并将项目名称设置为 tourism。在命令提示符窗口中，输入以下命令：

```
vue create tourism
```

按 Enter 键，选择 Manually select features，如图 10.2 所示。

图 10.2　选择 Manually select features

然后，按 Enter 键，选择 Router 和 CSS Pre-processors 选项，如图 10.3 所示。

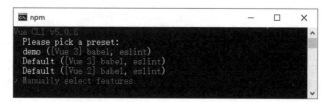

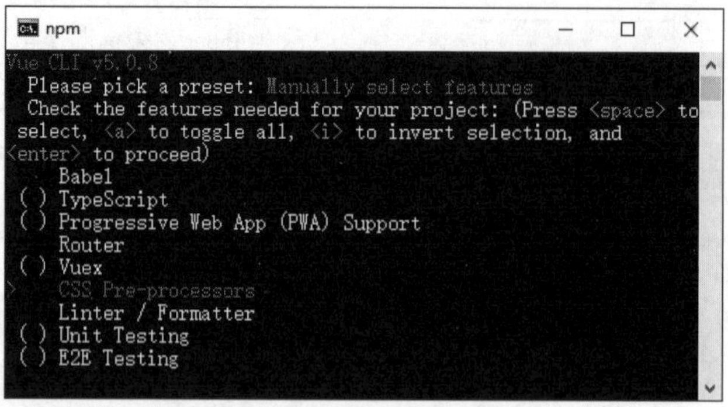

图 10.3　选择配置选项

再按 Enter 键，选择 Vue 的版本，这里选择 Vue 3.x 版本，如图 10.4 所示。

接着，选择路由是否使用 history 模式，输入 y 表示使用 history 模式，如图 10.5 所示。

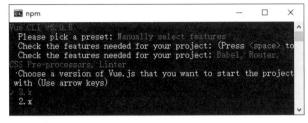 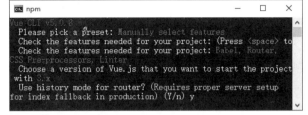

图 10.4　选择 Vue 版本　　　　　　　　图 10.5　使用 history 模式

在选择 CSS 预处理器时，选择 Sass/SCSS 选项，如图 10.6 所示。

在选择配置信息的存放位置时，选择 In package.json 选项，即将配置信息存储在 package.json 文件中，如图 10.7 所示。

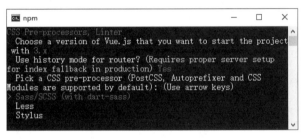 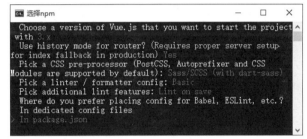

图 10.6　选择 CSS 预处理器　　　　　　图 10.7　选择配置信息的存放位置

创建项目后，进入项目目录，分别安装项目需要使用的插件和库，包括 axios、ElementPlus 和 Day.js。安装完成后，整理项目目录，在 public 目录中创建 img 文件夹，用于存储项目所需的图片文件。接着，在 assets 目录中创建 css 文件夹和 img 文件夹，分别用于存储项目所需的 CSS 文件和图片文件。

除此之外，因为在该项目中使用了 JSON Server 模拟后端服务器，所以还需要安装 JSON Server，在项目根目录下创建一个 scenery.json 文件作为数据源，在该文件中添加模拟服务器的数据。

10.5　公共组件设计

在开发项目时，编写公共组件可以减少重复代码的编写，有利于代码的重用和维护。在设计四季旅游信息网时，大部分页面都会使用两个公共组件：一个是页面头部组件 MyHeader.vue，另一个是页面底部组件 MyFooter.vue。下面将详细介绍这两个组件。

10.5.1　页面头部组件设计

页面头部组件主要提供网站导航栏的功能，其界面效果如图 10.8 所示。

图 10.8　页面头部组件

头部组件的实现过程比较简单。在 components 文件夹下新建 MyHeader.vue 文件，然后在<template>标签中添加一个 ul 列表，接着在该列表中使用<router-link>组件设置导航链接，并将<router-link>渲染为<li>标签，最后引入该组件使用的样式文件 nav.css。代码如下：

```vue
<template>
  <div class="nav my-header">
    <div class="center">
      <div class="area">
        <div>
          <img src="/img/logo.png" />
        </div>
        <ul>
          <router-link to="/" custom v-slot="{navigate, isExactActive}">
            <li :class="[isExactActive && 'router-link-exact-active']" @click="navigate">
              <div></div>
              首页
            </li>
          </router-link>
          <router-link to="/scenery_list" custom v-slot="{navigate, isExactActive}">
            <li :class="[isExactActive && 'router-link-exact-active']" @click="navigate">
              <div></div>
              热门景点
            </li>
          </router-link>
          <router-link to="/hotel" custom v-slot="{navigate, isExactActive}">
            <li :class="[isExactActive && 'router-link-exact-active']" @click="navigate">
              <div></div>
              酒店住宿
            </li>
          </router-link>
          <router-link to="/ticket" custom v-slot="{navigate, isExactActive}">
            <li :class="[isExactActive && 'router-link-exact-active']" @click="navigate">
              <div></div>
              门票预定
            </li>
          </router-link>
          <router-link to="/service" custom v-slot="{navigate, isActive}">
            <li :class="[isActive && 'router-link-active']" @click="navigate">
              <div></div>
              游客服务
            </li>
          </router-link>
          <router-link to="/user" custom v-slot="{navigate, isActive}">
            <li :class="[isActive && 'router-link-active']" @click="navigate">
              <div></div>
              用户中心
            </li>
          </router-link>
        </ul>
      </div>
    </div>
  </div>
</template>
<style scoped src="@/assets/css/nav.css"></style>
```

10.5.2 页面底部组件设计

页面底部组件主要展示友情链接、旅游资讯投诉热线和网站官方的联系方式，其界面效果如图 10.9 所示。

图 10.9 页面底部组件

底部组件的实现过程也比较简单。在 components 文件夹下新建 MyFooter.vue 文件，在<template>标签中使用<div>标签分别定义友情链接、旅游资讯投诉热线和网站官方的联系方式，最后引入该组件使用的样式文件 footer.css。代码如下：

```
<template>
  <div class="foot my-footer">
    <div class="center">
      <div class="main">
        <div class="box1">
          <p>友情链接</p>
          <div>
            <a href="#">明日旅游网</a>
            <a href="https://www.ctrip.com/">携程旅游网</a>
            <a href="https://www.tuniu.com/">途牛旅游网</a>
          </div>
        </div>
        <div class="box2">
          <img src="/img/icon1.png" />
          <span>旅游咨询投诉热线：12345</span>
        </div>
        <div class="box3">
          <div>
            <img src="/img/weibo.png" />
            <p>官方微博</p>
          </div>
          <div>
            <img src="/img/weixin.png" />
            <p>官方微信</p>
          </div>
        </div>
      </div>
    </div>
  </div>
</template>
<style scoped src="@/assets/css/footer.css"></style>
```

10.6 首页设计

四季旅游信息网的首页主要提供网站导航、旅游广告轮播图、热门景点图片展示和酒店住宿图片展示等内容。首页效果如图 10.10 所示。

图 10.10 首页

四季旅游信息网首页的实现过程如下。

（1）在 views 文件夹下新建 Home 文件夹，并在 Home 文件夹下新建首页组件 index.vue。接着，在 <template> 标签中分别定义旅游广告轮播图、热门景点图片列表、酒店住宿图片列表和无间断向左滚动的旅游景点图片列表。单击热门景点图片列表区域中的"更多"超链接，可以进入热门景点页面。同样，单击酒店住宿图片列表区域中的"更多"超链接，可以进入酒店住宿页面。代码如下：

```
<template>
  <div>
    <!-- 轮播图 -->
    <div id="box">
      <ul id="imagesUI" class="list">
        <transition-group name="fade" tag="div">
```

```html
          <li v-for="(v,i) in banners" :key="v" v-show="(i+1)===index?true:false"><img :src="v"></li>
        </transition-group>
      </ul>
      <ul id="btnUI" class="count">
        <li v-for="num in 5" :key="num" @mouseover='change(num)' :class='{current:num===index}'>
          {{num}}
        </li>
      </ul>
    </div>
    <!-- 热门景点 -->
    <div class="scenery">
      <div class="center">
        <div class="row1">
          <div class="txt2">
            <div class="icon"></div>
            <span>热门景点</span>
          </div>
          <router-link to="/scenery_list">更多&gt;&gt;</router-link>
        </div>
        <div class="pic">
          <div class="center">
            <div class="left-pic">
              <a href="/scenery_show?scenery_id=1"
                ><img src="img/index/jyt.jpg"
              /></a>
              <p>净月潭国家森林公园</p>
            </div>
            <div class="right-pic">
              <ul>
                <li>
                  <img src="/img/index/1.jpg" alt="" />
                  <p>森林浴场</p>
                </li>
                <li>
                  <img src="/img/index/2.jpg" alt="" />
                  <p>净月湿地</p>
                </li>
                <li>
                  <img src="/img/index/3.jpg" alt="" />
                  <p>净月雪世界</p>
                </li>
                <li>
                  <img src="/img/index/4.jpg" alt="" />
                  <p>净月女神</p>
                </li>
              </ul>
            </div>
          </div>
        </div>
      </div>
    </div>
    <div class="hotel">
      <div class="center">
        <div class="row1">
          <div class="txt2">
            <div class="icon"></div>
            <span>酒店住宿</span>
          </div>
          <router-link to="/hotel">更多&gt;&gt;</router-link>
        </div>
        <div class="pic">
          <div class="center">
```

```html
                <ul>
                    <li>
                        <img src="img/hotel/1.jpg" alt="" />
                        <p>益田喜来登酒店</p>
                    </li>
                    <li>
                        <img src="img/hotel/7.jpg" alt="" />
                        <p>华天大酒店</p>
                    </li>
                    <li>
                        <img src="img/hotel/9.jpg" alt="" />
                        <p>亚朵酒店</p>
                    </li>
                </ul>
            </div>
        </div>
    </div>
</div>
<div class="scenery2">
    <div class="center">
        <div class="txt2">
            <div class="icon"></div>
            <span>欢迎来长春旅游</span>
        </div>
        <div class="area">
            <ul>
                <li class="active">
                    <img src="img/scenery/1.jpg" />
                    <p>净月潭国家森林公园</p>
                </li>
                <li>
                    <img src="img/scenery/3.jpg" />
                    <p>动植物公园</p>
                </li>
                <li>
                    <img src="img/scenery/5.jpg" />
                    <p>世界雕塑园</p>
                </li>
                <li>
                    <img src="img/scenery/7.jpg" />
                    <p>东北虎园</p>
                </li>
                <li>
                    <img src="img/scenery/1.jpg" />
                    <p>净月潭国家森林公园</p>
                </li>
                <li>
                    <img src="img/scenery/3.jpg" />
                    <p>动植物公园</p>
                </li>
                <li>
                    <img src="img/scenery/5.jpg" />
                    <p>世界雕塑园</p>
                </li>
                <li>
                    <img src="img/scenery/7.jpg" />
                    <p>东北虎园</p>
                </li>
            </ul>
        </div>
    </div>
</div>
```

```
        </div>
</template>
```

（2）在<script>标签中首先定义与轮播图相关的数据，然后在 methods 选项中定义方法。其中，next()方法用于设置轮播图中图片的索引，change()方法在用户单击轮播图中的数字按钮时被调用，该方法用于将轮播图切换为数字按钮对应的图片。最后在 mounted 选项中设置每隔 3 秒自动切换一张图片。代码如下：

```
<script>
export default {
  name: 'TheIndex',
  data() {
    return {
      banners : [                                        //广告图片数组
        require('/public/img/banner/banner_1.jpg'),
        require('/public/img/banner/banner_2.jpg'),
        require('/public/img/banner/banner_3.jpg'),
        require('/public/img/banner/banner_4.jpg'),
        require('/public/img/banner/banner_5.jpg')
      ],
      index : 1,                                         //图片的索引
      flag : true,                                       //控制不允许连续单击数字按钮
      timer : '',                                        //定时器 ID
    }
  },
  methods : {
    next : function(){
      //下一张图片，图片索引为 5 时返回第一张
      this.index = this.index + 1 === 6 ? 1 : this.index + 1;
    },
    change : function(num){
      //鼠标移入按钮切换到对应图片
      if(this.flag){
        this.flag = false;
        //过 1 秒后可以再次移入按钮切换图片
        setTimeout(()=>{
          this.flag = true;
        },1000);
        this.index = num;                                //切换为选中的图片
        clearInterval(this.timer);                       //取消定时器
        //过 3 秒图片轮换
        this.timer = setInterval(this.next,3000);
      }
    }
  },
  mounted : function(){
    //过 3 秒图片轮换
    this.timer = setInterval(this.next,3000);
  },
  beforeUnmount : function(){
    clearInterval(this.timer);
  }
};
</script>
```

10.7 热门景点页面设计

单击导航栏中的"热门景点"超链接，或单击首页中热门景点图片列表区域的"更多"超链接，用户

可以进入热门景点页面。热门景点页面的效果如图 10.11 所示。

图 10.11　热门景点页面

热门景点页面的相关组件主要有 3 个，分别是景点列表组件、景点列表项组件和景点详情组件，下面对这 3 个组件进行详细介绍。

10.7.1　景点列表组件设计

景点列表组件主要包括季节分类、景点列表和分页部件。每页显示 4 个景点列表项，每个景点列表项都展示了景点图片、景点名称和景点门票价格等信息。景点列表组件的实现过程如下。

（1）在 views 文件夹下新建 Scenery 文件夹，并在 Scenery 文件夹下新建景点列表组件 SceneryList.vue。在<template>标签中：首先定义季节分类，利用 active 属性值的变化控制季节被单击后的样式；然后对景点列表进行遍历，根据 seasonId 的值判断遍历的是全部景点列表还是按季节分类的景点列表；在遍历景点列表的下面定义分页部件，根据 now 的值控制当前的页数。代码如下：

```
<template>
  <div class="body">
    <div class="center">
      <div class="txt" style="height: 10px;"></div>
    </div>
    <div class="search">
      <div class="center">
        <div class="main">
```

```html
            <div class="season">
              <span>季节：</span>
              <a :class="{active:active===0}" @click="seasonId = null,active = 0">不限</a>
              <a :class="{active:active===1}" @click="seasonId = 10,active = 1">春季游</a>
              <a :class="{active:active===2}" @click="seasonId = 20,active = 2">夏季游</a>
              <a :class="{active:active===3}" @click="seasonId = 30,active = 3">秋季游</a>
              <a :class="{active:active===4}" @click="seasonId = 40,active = 4">冬季游</a>
            </div>
          </div>
        </div>
      </div>
      <div class="scenery">
        <div class="center">
          <div class="main">
            <ul v-if="!seasonId">
              <scenery-cell v-for="scenery in sceneries" :key="scenery.scenery_id" :s="scenery"/>
            </ul>
            <ul v-if="seasonId">
              <scenery-cell v-for="scenery in sceneriesBySeason" :key="scenery.scenery_id" :s="scenery"/>
            </ul>
          </div>
        </div>
      </div>
      <!-- 分页器 -->
      <div class="center" v-show="pages">
        <div class="pages">
          <span :class="{disabled:now===1}" @click="now--">上一页</span>
          <span @click="now=n" :class="{active:now===n}" v-for="n in pages" :key="n">{{n}}</span>
          <span :class="{disabled:now===pages}" @click="now++">下一页</span>
        </div>
      </div>
    </div>
</template>
```

（2）在<script>标签中：首先分别引入 axios 实例和景点列表项组件 SceneryCell.vue；然后注册 SceneryCell 组件、定义数据，并对 seasonId 属性和 now 属性进行监听，根据属性值的变化执行不同的查询操作；接下来在 mounted()钩子函数中调用分页查询景点的方法 getSceneries()；最后定义 getSceneriesBySeason()方法和 getSceneries()方法，其中 getSceneriesBySeason()方法用于按照指定的季节查询景点，getSceneries()方法用于分页查询所有景点。代码如下：

```
<script>
import api from "@/utils/request";                              //引入 axios 实例
import SceneryCell from '@/components/SceneryCell.vue';
export default {
  components: { SceneryCell },
  data() {
    return {
      active:0,                                                  //激活当前单击季节的样式
      now: 1,                                                    //当前页码
      sceneries:[],                                              //分页查询热门景点的数据
      sceneriesBySeason:[],                                      //按照季节获取的景点
      seasonId: null,                                            //季节分类 id
      limit: 4,                                                  //每页显示数据条数
      pages: 0                                                   //总页数
    }
  },
  watch: {
    seasonId(newValue){
      if(newValue !== null) {                                    //如果单击了季节分类，就根据季节查询景点
        this.getSceneriesBySeason();
      } else {
```

```
            this.now = 1;
            this.getSceneries();
          }
        },
        now() {
          this.getSceneries()
        }
      },
      mounted () {
        this.getSceneries()
      },
      methods: {
        //按照季节查询热门景点
        async getSceneriesBySeason(){
          this.pages = 0;
          await api.get(`/list?seasonId=${this.seasonId}&_page=${this.now}&_per_page=${this.limit}`).then(res=>{
            this.sceneriesBySeason = res.data.data
          })
        },
        //分页查询所有景点
        async getSceneries() {
          await api.get('/list').then(function(res){
            this.pages = Math.ceil(res.data.length / this.limit);
          }.bind(this))
          await api.get(`/list?_page=${this.now}&_per_page=${this.limit}`).then(res=>{
            this.sceneries = res.data.data
          })
        }
      },
    };
</script>
```

10.7.2 景点列表项组件设计

景点列表项组件是景点列表组件的子组件。单个景点列表项的界面效果如图 10.12 所示。

图 10.12 景点列表项

景点列表项组件的实现过程比较简单。在 components 文件夹下新建景点列表项组件 SceneryCell.vue。首先在<template>标签中定义<li>列表项标签，并在<li>列表项标签中分别显示景点图片、景点名称和景点门票价格，单击景点图片会跳转到景点详情页面。然后在<script>标签中定义父组件传递的 Prop 属性。最后引入该组件使用的样式文件 sceneryList.css。代码如下：

```
<template>
  <li class="scenery-cell">
    <router-link :to="`/scenery_show?scenery_id=${s.scenery_id}`">
      <img :src="`${s.pic}`" />
    </router-link>
    <p>{{s.scenery_name}}</p>
```

```
      <span> ¥ {{s.scenery_price}}</span>
    </li>
</template>
<script>
export default {
  props:['s'],
};
</script>
<style scoped src="@/assets/css/sceneryList.css"></style>
```

10.7.3 景点详情组件设计

单击热门景点页面中的景点图片可以进入该景点的景点详情页面。该页面主要展示了景点图片、景点地理位置、景点开放时间、景点级别、门票价格和景点介绍等内容，效果如图10.13所示。

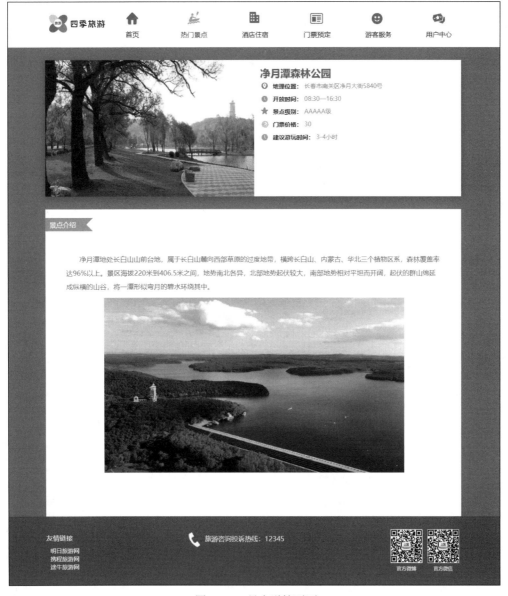

图 10.13　景点详情页面

景点详情组件的实现过程如下。

（1）在 Scenery 文件夹下新建景点详情组件 SceneryShow.vue。在<template>标签中定义<div>标签，并在<div>标签中分别定义景点图片、景点名称、景点地理位置、景点开放时间、景点级别、门票价格、建议游玩时间和景点介绍等内容。代码如下：

```
<template>
  <div class="body">
    <div class="center">
      <div class="txt" style="height:30px;"></div>
    </div>
    <div class="show">
      <div class="center">
        <div class="col1-1">
          <img :src="`${data[0]&&data[0].area_pic}`" />
        </div>
        <div class="col1-2">
          <div>
            <h3>{{data[0]&&data[0].scenery_name}}</h3>
          </div>
          <div>
            <div class="icon1"></div>
            <h4>地理位置：</h4>
            <p>{{data[0]&&data[0].address}}</p>
          </div>
          <div>
            <div class="icon2"></div>
            <h4>开放时间：</h4>
            <p>{{data[0]&&data[0].start_time}}—{{data[0]&&data[0].end_time}}</p>
          </div>
          <div>
            <div class="icon3"></div>
            <h4>景点级别：</h4>
            <p>{{data[0]&&data[0].level}}</p>
          </div>
          <div>
            <div class="icon4"></div>
            <h4>门票价格：</h4>
            <p>
              {{data[0]&&data[0].tickets}}
            </p>
          </div>
          <div>
            <div class="icon5"></div>
            <h4>建议游玩时间：</h4>
            <p>
              {{data[0]&&data[0].playtime}}
            </p>
          </div>
        </div>
      </div>
    </div>
    <div class="f2"></div>
    <div class="introduce">
      <div class="center">
        <div class="main">
          <div>
            <span>景点介绍</span>
            <span></span>
          </div>
          <div>
            <p>
```

```
            {{data[0]&&data[0].intro}}
          </p>
          <img :src="`${data[0]&&data[0].spot_pic}`"/>
          <br />
          <br />
        </div>
      </div>
    </div>
  </div>
</template>
```

（2）在\<script\>标签中首先引入 axios 实例；然后在 data 选项中定义用于存储景点详情的数据，在 mounted()钩子函数中调用 getData()方法；最后在 methods 选项中定义 getData()方法，该方法会根据景点 id 查询景点信息。代码如下：

```
<script>
import api from "@/utils/request";                        //引入 axios 实例
export default {
  data() {
    return {
      data: []                                            //景点详情
    };
  },
  mounted() {
    this.getData();
  },
  methods: {
    //根据景点 id 查询景点信息
    async getData() {
      await api.get(`/show?scenery_id=${this.$route.query.scenery_id}`).then((res) => {
        this.data = res.data;
      });
    },
  },
};
</script>
```

（3）在\<style\>标签中引入该组件使用的样式文件 sceneryShow.css。代码如下：

```
<style scoped src="@/assets/css/sceneryShow.css"></style>
```

10.8　酒店住宿页面设计

单击导航栏中的"酒店住宿"超链接，或单击首页中酒店住宿图片列表区域的"更多"超链接可以进入酒店住宿页面。酒店住宿页面的效果如图 10.14 所示。

酒店住宿页面的相关组件主要有 4 个，分别是酒店列表组件、酒店列表项组件、酒店搜索结果组件和酒店详情组件，下面对这 4 个组件进行详细介绍。

10.8.1　酒店列表组件设计

酒店列表组件主要包括酒店搜索框、酒店类型、选择每页显示数据条数的下拉菜单、酒店列表和分页部件。在默认情况下，每页显示 8 个酒店列表项，每个酒店列表项都展示酒店图片、酒店名称和酒店起始价格等信息。酒店列表组件的实现过程如下。

图10.14 酒店住宿页面

（1）在 views 文件夹下新建 Hotel 文件夹，在 Hotel 文件夹下新建酒店列表组件 HotelList.vue。在 <template>标签中：首先定义酒店搜索框和"搜索"按钮，单击"搜索"按钮调用 search()方法；然后定义酒店类型和用于选择每页显示数据条数的下拉菜单；接下来对酒店列表进行遍历，根据 hotelTypeId 的值判断遍历的是全部酒店列表还是指定类型的酒店列表；最后定义分页部件，根据 now 的值控制当前的页数。代码如下：

```
<template>
  <div class="body">
    <div class="search">
      <div class="center">
        <div class="btn-search">
          <h3>酒店住宿</h3>
          <div>
            <input type="text" v-model="keyword" placeholder="请输入酒店关键词" />
            <button @click="search">搜索</button>
          </div>
        </div>
      </div>
      <div class="main">
        <div class="hotel-type">
          <span>酒店类型：</span>
          <a :class="{active:active===0}" @click="hotelTypeId=null, active=0">不限</a>
          <a :class="{active:active===1}" @click="hotelTypeId=1, active=1">经济型</a>
```

```html
            <a :class="{active:active===2}" @click="hotelTypeId=2, active=2">舒适型</a>
            <a :class="{active:active===3}" @click="hotelTypeId=3, active=3">豪华型</a>
          </div>
        </div>
      </div>
    </div>
    <div class="hotel-list">
      <div class="center">
        <select class="number" v-model="limit" v-if="pages > 0">
          <option value="4">每页显示 4 条数据</option>
          <option value="8">每页显示 8 条数据</option>
          <option value="16">每页显示 16 条数据</option>
        </select>
        <ul v-if='!hotelTypeId'>
          <hotel-cell v-for="item in hotels" :key="item.hid" :item="item"/>
        </ul>
        <ul v-if="hotelTypeId">
          <hotel-cell v-for="item in hotelsByType" :key="item.hid" :item="item"/>
        </ul>
      </div>
    </div>
    <!-- 分页器 -->
    <div class="center" v-show="pages">
      <div class="pages">
        <span :class="{disabled:now===1}" @click="now--">上一页</span>
        <span @click="now=n" :class="{active:now===n}" v-for="n in pages" :key="n">{{n}}</span>
        <span :class="{disabled:now===pages}" @click="now++">下一页</span>
      </div>
    </div>
  </div>
</template>
```

（2）在<script>标签中：首先分别引入 axios 实例和酒店列表项组件 HotelCell.vue；然后注册 HotelCell 组件、定义数据，在 mounted()钩子函数中调用分页查询酒店的方法 getHotels()；接下来对 hotelTypeId 属性、limit 属性和 now 属性进行监听，根据属性值的变化执行相应的查询操作；最后在 methods 选项中分别定义 search()方法、getHotels()方法和 getHotelsByType()方法。其中：search()方法用于跳转到酒店搜索结果页面，在路由跳转时将搜索关键词作为参数进行传递；getHotels()方法用于查询所有酒店；getHotelsByType()方法用于按照酒店类型查询酒店。代码如下：

```
<script>
import api from "@/utils/request";                    //引入 axios 实例
import HotelCell from '@/components/HotelCell.vue';
export default {
  components: { HotelCell },
  data() {
    return {
      keyword: "",                                    //搜索关键词
      hotelsByType: [],                               //根据酒店分类查询的酒店数据
      hotelTypeId: null,                              //酒店分类
      hotels: [],                                     //所有酒店数据
      active: 0,                                      //激活当前单击酒店分类的样式
      now: 1,                                         //当前页码
      limit: 8,                                       //每页显示数据条数
      pages: 0                                        //总页数
    }
  },
  mounted () {
    this.getHotels()
  },
  watch: {
```

```
    hotelTypeId(value) {
      if(value !== null) {                    //如果单击了酒店分类,就隐藏分页功能
        this.pages = 0;
        this.getHotelsByType();
      } else {
        this.getHotels();
      }
    },
    limit(){
      this.getHotels();
    },
    now() {
      this.getHotels();
    }
  },
  methods: {
    search(){
      this.$router.push({
        path: "/search_result",
        query: {
          keyword: this.keyword
        }
      })
    },
    //查询所有酒店
    async getHotels(){
      await api.get('/hotel').then(function(res){
        this.pages = Math.ceil(res.data.length / this.limit);
      }.bind(this))
      await api.get(`/hotel?_page=${this.now}&_per_page=${this.limit}`).then(res=>{
        this.hotels = res.data.data;
      })
    },
    //按酒店类型查询酒店
    async getHotelsByType() {
      await api.get(`/hotel?hotel_type_id=${this.hotelTypeId} `).then(res=>{
        this.hotelsByType = res.data
      })
    }
  },
};
</script>
```

10.8.2 酒店列表项组件设计

酒店列表项组件是酒店列表组件的子组件。单个酒店列表项的界面效果如图 10.15 所示。

酒店列表项组件的实现过程比较简单。在 components 文件夹下新建酒店列表项组件 HotelCell.vue。首先在<template>标签中定义<li>列表项标签,在<li>列表项标签中分别显示酒店图片、酒店名称和酒店起始价格,单击酒店图片会跳转到酒店详情页面。然后在<script>标签中定义父组件传递的 Prop 属性,最后引入该组件使用的样式文件 hotel.css。代码如下:

```
<template>
  <li class="hotel-cell">
    <img :src="`${item.hotel_pic}`" />
    <router-link :to="`/hotel_show?hid=${item.hid}`">{{item.hotel_name}}</router-link>
    <h4>￥ {{item.price}}起</h4>
```

图 10.15 酒店列表项

```
    </li>
  </template>
<script>
export default {
  props:['item']
};
</script>
<style scoped src="@/assets/css/hotel.css"></style>
```

10.8.3 酒店搜索结果组件设计

在酒店住宿页面中的搜索框中输入关键词,单击"搜索"按钮后会跳转到酒店搜索结果页面,该页面会展示酒店名称中包含搜索关键词的所有酒店列表。例如,搜索酒店名称中包含"天"的所有酒店,搜索结果如图10.16所示。

酒店搜索结果组件的实现过程如下。

(1)在Hotel文件夹下新建酒店搜索结果组件SearchResult.vue。在<template>标签中对搜索结果进行遍历,将遍历的每一项作为Prop属性传递给酒店列表项组件。代码如下:

图10.16 酒店搜索结果

```
<template>
  <div class="body">
    <div class="search">
      <div class="center">
        <div class="btn-search">
          <h3>酒店住宿</h3>
        </div>
      </div>
    </div>
    <div class="hotel-list">
      <div class="center">
        <ul>
          <hotel-cell v-for="item in hotels" :key="item.hid" :item="item"/>
        </ul>
      </div>
    </div>
  </div>
</template>
```

(2)在<script>标签中:首先分别引入axios实例和酒店列表项组件HotelCell.vue;然后注册HotelCell组件、定义数据,在 mounted 钩子函数中调用查询酒店的方法 getHotels();最后在 methods 选项中定义getHotels()方法,该方法用于查询酒店名称中包含搜索关键词的所有酒店。代码如下:

```
<script>
import api from "@/utils/request";            //引入axios实例
import HotelCell from '@/components/HotelCell.vue';
export default {
  components: { HotelCell },
  data() {
    return {
      keyword: this.$route.query.keyword,     //搜索关键词
      hotels: []                              //所有酒店数据
    }
  },
  mounted () {
    this.getHotels();
  },
```

```
methods: {
    async getHotels(){
        await api.get('/hotel').then(res => {                    //查询所有酒店
            this.hotels = res.data;
            //获取酒店名称中包含搜索关键词的所有酒店
            this.hotels = this.hotels.filter(item => {
                return item.hotel_name.includes(this.keyword);
            })
        })
    }
};
</script>
```

10.8.4 酒店详情组件设计

单击酒店住宿页面中的酒店名称可以进入该酒店的酒店详情页面。该页面主要展示酒店图片、酒店名称、酒店起始价格、酒店地址、联系电话、入住时间、酒店介绍、酒店设施等内容，效果如图10.17所示。

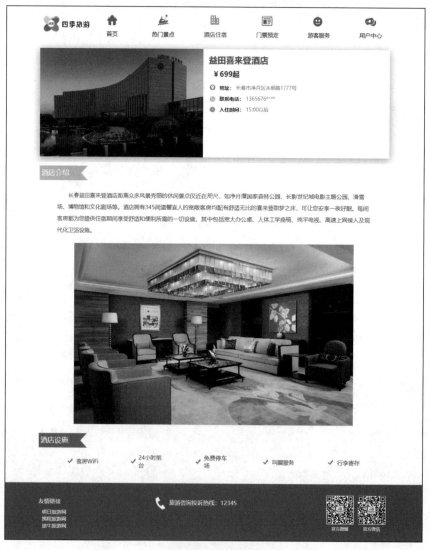

图 10.17 酒店详情页面

酒店详情组件的实现过程如下。

（1）在 Hotel 文件夹下新建酒店详情组件 HotelShow.vue。在<template>标签中定义<div>标签，在<div>标签中分别定义酒店图片、酒店名称、酒店起始价格、酒店地址、联系电话、入住时间、酒店介绍和酒店设施等内容。代码如下：

```
<template>
  <div>
    <div class="show" v-if="data">
      <div class="center">
        <div class="col1-1">
          <img :src="`${data[0]&&data[0].hotel_pic}`" />
        </div>
        <div class="col1-2">
          <div>
            <h3>{{data[0]&&data[0].hotel_name}}</h3>
          </div>
          <div class="price">
            <h3>¥{{data[0]&&data[0].price}}起</h3>
          </div>
          <div>
            <div class="icon1"></div>
            <h4>地址：</h4>
            <p>{{data[0]&&data[0].hotel_address}}</p>
          </div>
          <div>
            <div class="icon2"></div>
            <h4>联系电话：</h4>
            <p>{{data[0]&&data[0].hotel_phone}}</p>
          </div>
          <div>
            <div class="icon3"></div>
            <h4>入住时间：</h4>
            <p>{{data[0]&&data[0].checkin_time}}</p>
          </div>
        </div>
      </div>
    </div>
    <!-- 酒店介绍 -->
    <div class="introduce">
      <div class="center">
        <div class="icon">
          <span>酒店介绍</span>
          <div></div>
        </div>
        <p>
          {{data[0]&&data[0].intro}}
        </p>
        <img :src="`${data[0]&&data[0].room_pic}`" alt="" />
      </div>
    </div>
    <!-- 酒店设施 -->
    <div class="facility">
      <div class="center">
        <div class="icon">
          <span>酒店设施</span>
          <div></div>
        </div>
        <ul>
          <li>
            <div></div>
            <span>客房 WiFi</span>
```

```html
            </li>
            <li>
                <div></div>
                <span>24 小时前台</span>
            </li>
            <li>
                <div></div>
                <span>免费停车场</span>
            </li>
            <li>
                <div></div>
                <span>叫醒服务</span>
            </li>
            <li>
                <div></div>
                <span>行李寄存</span>
            </li>
        </ul>
      </div>
    </div>
</template>
```

（2）在<script>标签中：首先引入 axios 实例；然后在 data 选项中定义用于存储酒店详情的数据，在 mounted 钩子函数中调用 getData()方法；最后在 methods 选项中定义 getData()方法，该方法会根据酒店 id 查询酒店信息。代码如下：

```html
<script>
import api from "@/utils/request";                              //引入 axios 实例
export default {
  data() {
    return {
      data: []                                                  //酒店详情数据
    };
  },
  mounted() {
    this.getData();
  },
  methods: {
    //根据酒店 id 查询酒店信息
    async getData() {
      await api.get(`/hotel_detail?hid=${this.$route.query.hid}`).then((res) => {
        this.data = res.data;
      });
    },
  },
};
</script>
```

（3）在<style>标签中引入该组件使用的样式文件 hotelshow.css。代码如下：

```html
<style scoped src="@/assets/css/hotelshow.css"></style>
```

10.9　门票预订页面设计

用户如果未登录，单击导航栏中的"门票预订"超链接会跳转到用户登录页面；用户如果已登录，单击导航栏中的"门票预订"超链接可直接进入门票预订页面。该页面主要展示景点名称、可以选择的日期、门票购买张数、新增游客信息表单、购票须知和使用说明等内容，效果如图10.18所示。

图10.18 门票预订页面

门票预订页面的主要组件是门票预订组件TicketPage.vue，该组件的实现过程如下。

（1）在views文件夹下新建Ticket文件夹，并在Ticket文件夹下新建门票预订组件TicketPage.vue。在<template>标签中，首先定义景点名称。然后设置可以选择的日期，使用DatePicker组件定义日期选择器。接着定义"+"按钮和"-"按钮，以便用户单击这两个按钮来设置门票购买张数。再定义新增游客信息表单，该表单将要求用户输入姓名、手机号和身份证号。最后定义购票须知和门票使用说明等内容。代码如下：

```
<template>
  <div class="main body">
    <div class="center">
      <div class="left">
        <div class="f1">
          <p>净月潭风景名胜区门票（成人票）</p>
          <span>¥{{ unitPrice }}/张</span>
        </div>
        <div class="f2">
          <span>选择日期</span>
          <div class="date">
            <ul>
              <li @click="pick(1)" :class="{ active: now === 1 }">
                <span>今天({{ getDate(0) }})</span>
                <h4>¥30</h4>
              </li>
              <li @click="pick(2)" :class="{ active: now === 2 }">
                <span>明天({{getDate(1) }})</span>
                <h4>¥30</h4>
              </li>
              <li @click="pick(3)" :class="{ active: now === 3 }">
                <span>后天({{getDate(2) }})</span>
                <h4>¥30</h4>
```

```html
              </li>
              <li style="border:none">
                <el-date-picker
                    v-model="date"
                    format="订 M 月 D 日"
                    style="width:110px;font-size:12px"
                    type="date"
                    placeholder="其他日期"
                    @change="now=0"
                >
                </el-date-picker>
              </li>
          </ul>
        </div>
      </div>
      <div class="f3">
        <div>购买张数</div>
        <div>
          <button :disabled="number === 1" @click="number--">-</button>
          <input type="text" :value="number" />
          <button :disabled="number >= 10" @click="number++">+</button>
        </div>
        <div>最多订购 10 张</div>
      </div>
      <div class="user">
        <div class="user_f1">
          <div>新增游客信息</div>
          <div>请填写游客信息</div>
        </div>
        <div class="user_msg">
          <el-form :model="user" :rules="rules" ref="orderForm">
            <el-form-item class="ele" prop="tourist_name" label="姓名：" label-width="100px">
              <el-input v-model="user.tourist_name" placeholder="姓名"></el-input>
            </el-form-item>
            <el-form-item class="ele" prop="phone" label="手机号：" label-width="100px">
              <el-input type="password" v-model="user.phone" placeholder="手机号"></el-input>
            </el-form-item>
            <el-form-item class="ele" prop="identity_card" label="身份证号：" label-width="100px">
              <el-input v-model="user.identity_card" placeholder="身份证号"></el-input>
            </el-form-item>
          </el-form>
        </div>
      </div>
      <div class="total">
        <span>订单总价</span>
        <span>￥{{ unitPrice * number }}</span>
        <span>({{ number }}张)</span>
        <el-button type="primary" class="but" @click="touristAdd">提交订单</el-button>
      </div>
    </div>
    <div class="right">
      <div class="title">购票须知</div>
      <div class="refund">
        <h4>无须换票</h4>
        <p>无须换票，凭[入园码]直接入园，外籍游客持护照入园。</p><br>
        <h4>随时退</h4>
        <p>未使用可随时申请全额退款</p>
      </div>
      <div class="title">使用说明</div>
      <div class="msg">
        <h4>入园时间</h4>
        <p>08:30—16:30</p><br>
```

```
          <h4>入园地址</h4>
          <p>净月潭游客服务中心，净月潭风景名胜区正门</p>
        </div>
        <button><a href="#mask">详情购买须知&gt;&gt;</a></button>
      </div>
    </div>
  </div>
</template>
```

（2）在<script>标签中首先分别引入 Day.js、ElMessage 组件和 axios 实例，然后在 data 选项中分别定义验证手机号的规则、验证身份证号的规则、数据和表单验证规则。接下来在 methods 选项中定义方法。其中：pick()方法用于控制哪个日期的样式被激活，并设置日期选择器不显示日期；getDate()方法用于获取指定日期；touristAdd()方法用于判断表单验证是否成功，如果表单验证成功，就向 JSON 服务器中添加数据。最后在 mounted()钩子函数中判断用户是否已登录，如果用户未登录，就给出消息提示，并跳转到登录页面。代码如下：

```
<script>
import dayjs from "dayjs";                                 //引入 Day.js
import {ElMessage} from "element-plus";                    //引入 ElMessage 组件
import api from "@/utils/request";                         //引入 axios 实例
export default {
  data() {
    //验证手机号
    let validatorPhone = (rule, value, callback) => {
      if (!/^1[345789]\d{9}$/.test(this.user.phone)) {
        callback(new Error('请输入正确的手机号！'));
      } else {
        callback();
      }
    };
    //验证身份证号
    let validatorCard = (rule, value, callback) => {
      if (!(/(^\d{15}$)|(^\d{17}([0-9]|X)$)/.test(this.user.identity_card))) {
        callback(new Error('请输入正确的身份证号！'));
      } else {
        callback();
      }
    };
    return {
      unitPrice: 30,                                       //门票价格
      number: 1,                                           //购买张数
      user: {
        tourist_name: "",                                  //游客姓名
        phone: "",                                         //手机号
        identity_card: ""                                  //身份证
      },
      now: 1,                                              //日期激活
      date: "",                                            //日期选择器选择的日期
      //表单验证规则
      rules: {
        tourist_name: [
          { required: true, message: '请输入姓名', trigger: 'blur' },
          { min: 3, max: 15, message: '用户名为 3~15 个字符', trigger: 'blur' }
        ],
        phone: [
          { required: true, message: '请输入手机号', trigger: 'blur' },
          { validator: validatorPhone, trigger: 'blur' }
        ],
        identity_card: [
          { required: true, message: '请输入身份证号', trigger: 'blur' },
```

```
          { validator: validatorCard, trigger: ['blur', 'change'] }
        ]
      }
    };
  },
  methods: {
    pick(num){
      this.now = num;
      this.date = "";                                          //日期选择器不显示日期
    },
    //获取指定日期
    getDate(num){
      return dayjs().add(num, "day").format("M 月 DD 日")
    },
    async touristAdd() {
      await this.$refs.orderForm.validate((valid) => {         //验证表单元素
        if (valid) {                                           //如果验证成功,就添加数据
          api({
            method: "post",
            url: "/tourist",
            data: {
              tourist_name: this.user.tourist_name,
              identity_card: this.user.identity_card,
              phone: this.user.phone
            }
          }).then(() => {
            ElMessage({
              message: "订单提交成功! ",
              type: "success"
            })
            this.$refs.orderForm.resetFields();
          })
        }
      });
    }
  },
  mounted() {
    if(!sessionStorage.getItem("uname")){
      ElMessage({
        message: "请您先登录! ",
        type: "warning"
      })
      this.$router.push("/login");                             //跳转到登录页面
    }
  }
};
</script>
```

10.10 游客服务页面设计

单击导航栏中的"游客服务"超链接,用户可以进入游客服务页面。在游客服务页面中,用户可以查看导游和游客须知等信息。游客服务页面的效果如图 10.19 所示。

由于游客服务页面中的"交通查询"和"投诉热线"两个选项并没有实质上的功能,因此与游客服务页面相关的组件主要有 3 个,分别是游客服务组件、导游组件和游客须知组件。下面对这 3 个组件分别进行详细介绍。

四季旅游信息网 第 10 章

图 10.19 游客服务页面

10.10.1 游客服务组件设计

游客服务组件是一级路由组件，该组件主要提供"导游""游客须知""交通查询""投诉热线"4 个服务选项。限于本书篇幅，这里仅为"导游"和"游客须知"设置超链接，单击某个超链接可以进入对应的二级路由渲染的页面。游客服务组件的实现过程如下。

（1）在 views 文件夹下新建 Service 文件夹，在 Service 文件夹下新建游客服务组件 ServicePage.vue。在<template>标签中定义游客服务选项，包括"导游""游客须知""交通查询"和"投诉热线"。其中，为"导游"和"游客须知"两个选项设置路由跳转。在游客服务选项下面使用<router-view>渲染二级路由组件。代码如下：

```
<template>
  <div class="body">
    <div class="service_bg">
      <img src="/img/service/service.jpg" alt="" />
    </div>
    <div class="service">
      <div class="center">
        <div class="main">
          <div class="menu">
            <ul>
              <router-link to="/service/guide" custom v-slot="{navigate, isExactActive}">
                <li :class="[isExactActive && 'router-link-exact-active']" @click="navigate">
                  <div class="icon1"></div>
```

```
                        导游
                    </li>
                </router-link>
                <router-link to="/service/notice" custom v-slot="{navigate, isExactActive}">
                    <li :class="[isExactActive && 'router-link-exact-active']" @click="navigate">
                        <div class="icon2"></div>
                        游客须知
                    </li>
                </router-link>
                <li>
                    <div class="icon3"></div>
                    <p>交通查询</p>
                </li>
                <li>
                    <div class="icon4"></div>
                    <p>投诉热线</p>
                </li>
            </ul>
        </div>
        <div class="detail">
            <router-view/>
        </dlv>
    </div>
  </div>
 </div>
</template>
```

（2）在<style>标签中引入该组件使用的样式文件 service.css。代码如下：

```
<style scoped src="@/assets/css/service.css"></style>
```

10.10.2 导游组件设计

导游组件是二级路由渲染的组件。该组件主要提供导游信息，包括导游姓名、工作单位、工作经验年限和联系电话。导游组件的界面效果如图 10.20 所示。

图 10.20 导游组件

导游组件的实现过程比较简单。在 Service 文件夹下新建导游组件 ServiceGuide.vue。在<template>标签中对导游列表进行遍历，在遍历时输出导游头像、导游姓名、工作单位、工作经验年限和联系电话。然后在<script>标签中定义导游列表。最后引入该组件使用的样式文件 service.css。代码如下：

```html
<template>
  <div class="guide">
    <h4>导游</h4>
    <ul>
      <li v-for="item in guide" :key="item">
        <img :src="item.head" />
        <div class="introduce">
          <div>
            <h4>姓 名：</h4>
            <p>{{ item.name }}</p>
          </div>
          <div>
            <h4>单 位：</h4>
            <p>{{ item.unit }}</p>
          </div>
          <div>
            <h4>工作经验：</h4>
            <p>{{ item.years }}年</p>
          </div>
          <div>
            <h4>联系电话：</h4>
            <p>{{ item.phone }}</p>
          </div>
        </div>
      </li>
    </ul>
  </div>
</template>

<script>
export default {
  data(){
    return {
      guide: [
        {
          head: "/img/guide/1.jpg",
          name: "张三",
          unit: "青年旅行社有限公司",
          years: 5,
          phone: "136****2616"
        },
        {
          head: "/img/guide/1.jpg",
          name: "李四",
          unit: "通达旅行社",
          years: 3,
          phone: "1573526****"
        },
        {
          head: "/img/guide/1.jpg",
          name: "王五",
          unit: "文化国旅旅游有限公司",
          years: 6,
          phone: "132****5676"
        }
      ]
    }
  }
};
</script>
<style scoped src="@/assets/css/service.css"></style>
```

10.10.3 游客须知组件设计

游客须知组件也是二级路由渲染的组件。该组件主要提供游客须知信息，包括旅游时的注意事项和温馨提示内容。游客须知组件的界面效果如图 10.21 所示。

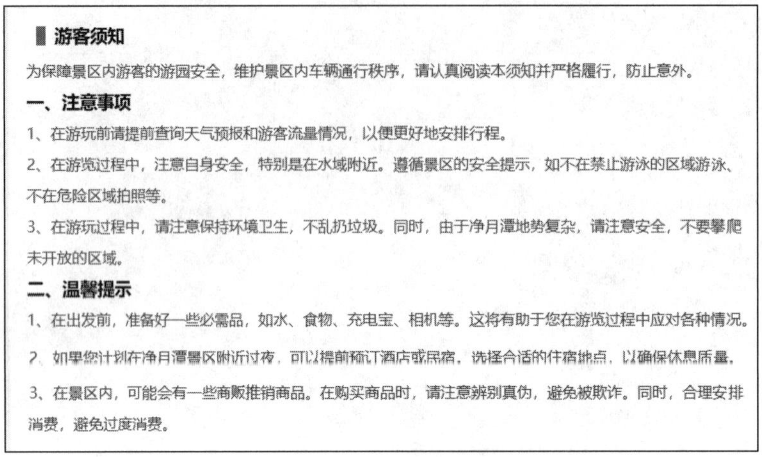

图 10.21 游客须知组件

游客须知组件是 ServiceNotice.vue。在 Service 文件夹下新建该组件，在<template>标签中定义游客须知信息，包括旅游时的注意事项和温馨提示。代码如下：

```
<template>
  <div class="notice">
    <div class="txt">
      <span></span>
      <h3>游客须知</h3>
    </div>
    <p>
      为保障景区内游客的游园安全，维护景区内车辆通行秩序，请认真阅读本须知并严格履行，防止意外。
    </p>
    <h3>一、注意事项</h3>
    <p>
      1、在游玩前请提前查询天气预报和游客流量情况，以便更好地安排行程。
    </p>
    <p>2、在游览过程中，注意自身安全，特别是在水域附近。遵循景区的安全提示，如不在禁止游泳的区域游泳、不在危险区域拍照等。</p>
    <p>
      3、在游玩过程中，请注意保持环境卫生，不乱扔垃圾。同时，由于净月潭地势复杂，请注意安全，不要攀爬未开放的区域。
    </p>
    <h3>二、温馨提示</h3>
    <p>
      1、在出发前，准备好一些必需品，如水、食物、充电宝、相机等。这将有助于您在游览过程中应对各种情况。
    </p>
    <p>
      2、如果您计划在净月潭景区附近过夜，可以提前预订酒店或民宿。选择合适的住宿地点，以确保休息质量。
    </p>
    <p>
      3、在景区内，可能会有一些商贩推销商品。在购买商品时，请注意辨别真伪，避免被欺诈。同时，合理安排消费，避免过度消费。
    </p>
  </div>
</template>
```

10.11 用户中心页面设计

用户中心页面包括注册页面和登录页面。单击导航栏中的"用户中心"超链接，默认会进入用户登录页面，该页面效果如图 10.22 所示。用户如果还未注册，则需要先进行注册，才能进行登录。单击登录页面中的"立即注册"超链接会进入用户注册页面，该页面效果如图 10.23 所示。

图 10.22　用户登录页面

图 10.23　用户注册页面

用户中心页面的相关组件包括用户注册组件和用户登录组件，下面对这两个组件进行详细介绍。

10.11.1　用户注册组件设计

用户注册组件主要由注册表单组成。用户需要在注册表单中输入用户名、密码、确认密码和邮箱。在注册过程中，系统会对用户输入的内容进行验证：如果输入的内容不符合要求，系统会显示相应的提示信息，效果如图 10.24 所示；如果验证通过，用户单击"注册"按钮后，系统会提示注册成功。

图 10.24 注册表单验证效果

用户注册组件的实现过程如下。

（1）在 views 文件夹下新建 User 文件夹，在 User 文件夹下新建用户注册组件 UserRegister.vue。在 <template> 标签中使用 ElementPlus 中的 Form 组件、FormItem 组件、Input 组件和 Button 组件定义用户注册表单，并设置表单的验证规则。将输入框和定义的数据进行绑定。当用户单击"注册"按钮时，调用 submitForm() 方法；当用户单击"重置"按钮时，调用 resetForm() 方法。代码如下：

```
<template>
  <div class="body">
    <div class="center">
      <div style="height: 10px"></div>
      <div class="login">
        <div class="pic">
          <img src="/img/reg/2.jpg" class="pic1" />
        </div>
        <div class="msg">
          <div class="user-title">
            <img src="/img/login/user.png" alt="" />
            <span>注  册</span>
          </div>
          <div class="reg">
            <span>已有账户?</span>
            <router-link to="/user/login">立即登录</router-link>
          </div>
          <el-form :model="form" :rules="rules" ref="registerForm">
            <el-form-item class="ele" prop="username" label="用户名：" label-width="100px">
              <el-input v-model="form.username" placeholder="用户名"></el-input>
            </el-form-item>
            <el-form-item class="ele" prop="password" label="密码：" label-width="100px">
              <el-input type="password" v-model="form.password" placeholder="密码"></el-input>
            </el-form-item>
            <el-form-item class="ele" prop="rePassword" label="确认密码：" label-width="100px">
              <el-input type="password" v-model="form.rePassword" placeholder="确认密码"></el-input>
            </el-form-item>
            <el-form-item class="ele" prop="email" label="邮箱：" label-width="100px">
              <el-input v-model="form.email" placeholder="邮箱"></el-input>
            </el-form-item>
            <el-form-item class="ele">
              <el-button type="primary" style="width: 100px;" size="default" @click="submitForm">注册</el-button>
```

```
                <el-button type="info" style="width: 100px;" size="default" @click="resetForm">重置</el-button>
            </el-form-item>
          </el-form>
        </div>
      </div>
      <div style="height: 50px"></div>
    </div>
  </div>
</template>
```

（2）在<script>标签中分别引入 ElMessage 组件和 axios 实例，然后在 data 选项中分别定义验证密码的规则、验证确认密码的规则、数据和表单验证规则，接下来在 methods 选项中定义 submitForm()方法和 resetForm()方法。其中：submitForm()方法用于判断表单验证是否成功，如果表单验证成功，就向 JSON 服务器中添加用户注册信息，并跳转到登录页面；resetForm()方法用于重置表单。代码如下：

```
<script>
import {ElMessage} from "element-plus";                    //引入 ElMessage 组件
import api from "@/utils/request";                          //引入 axios 实例
export default {
  data() {
    //验证密码
    let validatorPass = (rule, value, callback) => {
      if (value.length < 6 || value.length > 15) {
        callback(new Error('密码为 6~15 个字符'));
      } else {
        if(this.form.rePassword !== ""){
          this.$refs.registerForm.validateField('rePassword', () => null)
        }
        callback();
      }
    };
    //验证确认密码
    let validatorRePass = (rule, value, callback) => {
      if (value !== this.form.password) {
        callback(new Error('两次输入密码不一致!'));
      } else {
        callback();
      }
    };
    return {
      form: {
        username: '',                                       //用户名
        password: '',                                       //密码
        rePassword: '',                                     //确认密码
        email: ''                                           //邮箱
      },
      //表单验证规则
      rules: {
        username: [
          { required: true, message: '请输入用户名', trigger: 'blur' },
          { min: 3, max: 15, message: '用户名为 3~15 个字符', trigger: 'blur' }
        ],
        password: [
          { required: true, message: '请输入密码', trigger: 'blur' },
          { validator: validatorPass, trigger: 'blur' }
        ],
        rePassword: [
          { required: true, message: '请输入确认密码', trigger: 'blur' },
          { validator: validatorRePass, trigger: 'blur' }
        ],
        email: [
```

```js
                    { required: true, message: '请输入邮箱', trigger: 'blur' },
                    { type: 'email', message: '请输入正确的邮箱地址', trigger: ['blur', 'change'] }
                ]
            }
        };
    },
    methods: {
        async submitForm() {
            await this.$refs.registerForm.validate((valid) => {           //验证表单元素
                if (valid) {                                               //如果验证成功，就添加数据
                    api({
                        method: "post",
                        url: "/user",
                        data: {
                            user_name: this.form.username,
                            user_pwd: this.form.password,
                            email: this.form.email
                        }
                    }).then(() => {
                        ElMessage({
                            message: "注册成功！",
                            type: "success"
                        })
                        this.$router.push("/login");                       //跳转到登录页面
                    });
                }
            });
        },
        resetForm(){
            this.$refs.registerForm.resetFields();                         //重置表单
        }
    },
};
</script>
```

（3）在<style>标签中引入该组件使用的样式文件 register.css。代码如下：

```html
<style scoped src="@/assets/css/register.css"></style>
```

10.11.2 用户登录组件设计

用户注册成功之后会跳转到登录页面，登录页面主要由登录表单组成。用户需要在登录表单中输入用户名和密码。在登录过程中，系统将验证输入框是否为空，效果如图 10.25 所示。如果输入内容不为空且输入的用户名和密码都正确，用户单击"登录"按钮后，系统会提示登录成功，并随即跳转到门票预订页面。

图 10.25 登录表单验证效果

用户登录组件的实现过程如下。

（1）在 User 文件夹下新建用户登录组件 UserLogin.vue。在<template>标签中，首先定义"立即注册"超链接，单击该超链接会跳转到用户注册页面。然后使用 ElementPlus 中的 Form 组件、FormItem 组件、Input 组件和 Button 组件定义用户登录表单，并设置表单的验证规则。将输入框和定义的数据进行绑定。当用户单击"登录"按钮时，调用 submitForm()方法；当用户单击"重置"按钮时，调用 resetForm()方法。代码如下：

```
<template>
  <div class="body">
    <div class="center">
      <div style="height: 10px"></div>
      <div class="login">
        <div class="pic">
          <img src="/img/login/2.jpg" class="pic1" />
        </div>
        <div class="msg">
          <div class="user-title">
            <img src="/img/login/user.png" alt="" />
            <span>登  录</span>
          </div>
          <div class="reg">
            <span>还没有账号?</span>
            <router-link to="/user/register">立即注册</router-link>
          </div>
          <el-form :model="form" :rules="rules" ref="loginForm">
            <el-form-item class="ele" prop="username" label="用户名：" label-width="100px">
              <el-input v-model="form.username" placeholder="用户名"></el-input>
            </el-form-item>
            <el-form-item class="ele" prop="password" label="密码：" label-width="100px">
              <el-input type="password" v-model="form.password" placeholder="密码"></el-input>
            </el-form-item>
            <el-form-item class="ele">
              <el-button type="primary" style="width: 100px;" @click="submitForm">登录</el-button>
              <el-button type="info" style="width: 100px;" @click="resetForm">重置</el-button>
            </el-form-item>
          </el-form>
        </div>
      </div>
      <div style="height: 50px"></div>
    </div>
  </div>
</template>
```

（2）在<script>标签中分别引入 ElMessage 组件和 axios 实例，然后在 data 选项中分别定义数据和表单验证规则，接下来在 methods 选项中定义 submitForm()方法和 resetForm()方法。其中，submitForm()方法用于判断表单验证是否成功。如果表单验证成功，该方法会从 JSON 服务器中查询用户名，如果用户名正确，该方法会继续判断输入的密码是否正确。如果密码也正确，系统将提示登录成功，并跳转到门票预订页面。resetForm()方法用于重置表单。代码如下：

```
<script>
import {ElMessage} from "element-plus";              //引入 ElMessage 组件
import api from "@/utils/request";                    //引入 axios 实例
export default {
  data() {
    return {
      form: {
        username: '',                                  //用户名
        password: '',                                  //密码
```

```
            },
            //表单验证规则
            rules: {
                username: [
                    { required: true, message: '请输入用户名', trigger: 'blur' }
                ],
                password: [
                    { required: true, message: '请输入密码', trigger: 'blur' }
                ]
            }
        };
    },
    methods: {
        async submitForm() {
            await this.$refs.loginForm.validate((valid) => {
                if(valid){
                    api.get(`/user?user_name=${this.form.username}`)            //查询用户名
                        .then(function (res) {
                            console.log(this.form.username)
                            if(res.data.length === 0){
                                ElMessage({
                                    message: "用户名不正确！",
                                    type: "error"
                                })
                            //判断密码是否正确
                            }else if(res.data[0].user_pwd !== this.form.password){
                                ElMessage({
                                    message: "登录密码不正确！",
                                    type: "error"
                                })
                            } else {
                                ElMessage({
                                    message: "登录成功！",
                                    type: "success"
                                })
                                sessionStorage.setItem("uname",this.form.username);  //保存用户名
                                this.$router.push('/ticket');                        //跳转到门票预订页面
                            }
                        }.bind(this));
                }
            });
        },
        resetForm(){
            this.$refs.loginForm.resetFields();                                 //重置表单
        }
    },
};
</script>
```

（3）在<style>标签中引入该组件使用的样式文件 login.css，代码如下：

```
<style scoped src="@/assets/css/login.css"></style>
```

10.12 路 由 配 置

下面给出该项目的路由配置，包括定义路由、创建路由对象、设置路由跳转后页面置顶，以及当路由发生变化时修改页面标题的功能。代码如下：

```js
import {createRouter,createWebHistory} from 'vue-router'
const routes = [
  {
    path: '/user',
    name: 'user',
    redirect: '/user/login',
    children: [
      {
        path: 'login',
        name: 'login',
        component: () => import('../views/User/UserLogin.vue'),
        meta: {
          isHideFooter: true,
          title: "用户登录"
        }
      },
      {
        path: 'register',
        name: 'register',
        component: () => import('../views/User/UserRegister.vue'),
        meta: {
          isHideFooter: true,
          title: "用户注册"
        }
      }
    ]
  },
  {
    path: '/service',
    name: 'service',
    component: () => import('../views/Service/ServicePage.vue'),
    children:[
      {
        path: 'notice',
        name: 'notice',
        component: () => import('../views/Service/ServiceNotice.vue'),
        meta: {
          title: "游客须知"
        }
      },
      {
        path: 'guide',
        name: 'guide',
        component: () => import('../views/Service/ServiceGuide.vue'),
        meta: {
          title: "导游"
        }
      },
      {
        path: '',
        component: () => import('../views/Service/ServiceGuide.vue'),
        redirect: '/service/guide'
      }
    ]
  },
  {
    path: '/ticket',
    name: 'ticket',
    component: () => import('../views/Ticket/TicketPage.vue'),
    meta: {
```

```
          title: "门票预订"
        }
      },
      {
        path: '/hotel_show',
        name: 'hotelShow',
        component: () => import('../views/Hotel/HotelShow.vue'),
        meta: {
          title: "酒店详情"
        }
      },
      {
        path: '/hotel',
        name: 'hotel',
        component: () => import('../views/Hotel/HotelList.vue'),
        meta: {
          title: "酒店住宿"
        }
      },
      {
        path: '/search_result',
        name: 'SearchResult',
        component: () => import('../views/Hotel/SearchResult.vue'),
        meta: {
          title: "酒店搜索结果"
        }
      },
      {
        path: '/scenery_show',
        name: 'sceneryShow',
        component: () => import('../views/Scenery/SceneryShow.vue'),
        meta: {
          title: "景点详情"
        }
      },
      {
        path: '/scenery_list',
        name: 'sceneryList',
        component: () => import('../views/Scenery/SceneryList.vue'),
        meta: {
          title: "热门景点"
        }
      },
      {
        path: '/',
        name: 'Index',
        component: () => import('../views/Home/index.vue'),
        meta: {
          title: "首页"
        }
      }
]
const router = createRouter({
  history: createWebHistory(process.env.BASE_URL),
  routes,
  //跳转页面后置顶
  scrollBehavior(to,from,savedPosition){
     if(savedPosition){
        return savedPosition;
     }else{
```

```
            return {
                top: 0,
                left: 0
            }
        }
    }
})
router.beforeEach((to, from, next) => {
    //路由发生变化时修改页面 title
    if (to.meta.title) {
        document.title = to.meta.title
    }
    next()
})
export default router
```

10.13 项目运行

通过前述步骤，我们已经设计并完成了"四季旅游信息网"项目的开发。接下来，我们运行该项目，以检验我们的开发成果。首先打开 package.json 文件，在 scripts 配置选项中添加一个脚本命令，添加后的代码如下：

```
"scripts": {
    "serve": "vue-cli-service serve",
    "build": "vue-cli-service build",
    "lint": "vue-cli-service lint",
    "mock": "json-server scenery.json --port 3000"
}
```

打开命令提示符窗口，切换到项目目录，执行启动 JSON Server 的命令，如图 10.26 所示。新打开一个命令提示符窗口，切换到项目所在目录，执行 npm run serve 命令运行该项目，如图 10.27 所示。

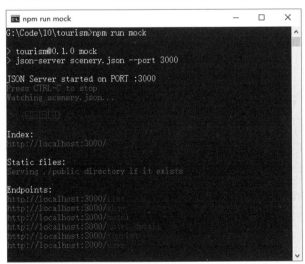

图 10.26 启动 JSON Server

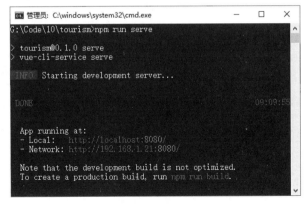

图 10.27 运行项目

在浏览器地址栏中输入 http://localhost:8080，然后按 Enter 键，即可进入四季旅游信息网的首页，效果如图 10.28 所示。

图 10.28 四季旅游信息网首页效果

10.14 源码下载

本章虽然详细地讲解了如何编码实现"四季旅游信息网"的各个功能，但给出的代码都是代码片段，而非完整的源代码。为了方便读者学习，本书提供了该项目的完整源代码，读者可以扫描右侧的二维码进行下载。

源码下载